SEISMIC PERFORMANCE OF SOIL-FOUNDATION-STRUCTURE SYSTEMS

SELECTED PAPERS FROM THE INTERNATIONAL WORKSHOP ON SEISMIC PERFORMANCE OF SOIL-FOUNDATION-STRUCTURE SYSTEMS, AUCKLAND, NEW ZEALAND, 21–22 NOVEMBER 2016

Seismic Performance of Soil-Foundation-Structure Systems

Editors

Nawawi Chouw, Rolando P. Orense & Tam Larkin
The University of Auckland, New Zealand

CRC Press
Taylor & Francis Group
Boca Raton   London   New York   Leiden

CRC Press is an imprint of the
Taylor & Francis Group, an **informa** business

A BALKEMA BOOK

Cover photo: Clustered structures in a laminar box by N. Chouw (University of Auckland)

CRC Press/Balkema is an imprint of the Taylor & Francis Group, an informa business

© 2017 Taylor & Francis Group, London, UK

Typeset by V Publishing Solutions Pvt Ltd., Chennai, India
Printed and Bound by CPI Group (UK) Ltd, Croydon, CR0 4YY

Published by: CRC Press/Balkema
             Schipholweg 107C, 2316 XC Leiden, The Netherlands
             e-mail: Pub.NL@taylorandfrancis.com
             www.crcpress.com – www.taylorandfrancis.com

ISBN: 978-1-138-06251-1 (Hbk)
ISBN: 978-1-315-16156-3 (eBook)

Seismic Performance of Soil-Foundation-Structure Systems – Chouw, Orense & Larkin (Eds)
© 2017 Taylor & Francis Group, London, ISBN 978-1-138-06251-1

Table of contents

Photos of workshop participants

Participants of the 2016 International Workshop on Seismic Performance of Soil-Foundation-Structure Systems

Participants of the 2016 International Workshop on Seismic Performance of Soil-Foundation-Structure Systems

Seismic Performance of Soil-Foundation-Structure Systems – Chouw, Orense & Larkin (Eds)
© 2017 Taylor & Francis Group, London, ISBN 978-1-138-06251-1

Preface

In conventional seismic design the concept of fixed base is deeply embedded in practice. This design approach is employed largely by virtue of its simplicity and this simplicity enables structural engineers to proceed with their design in isolation of geotechnical consideration. In this simple design approach the earthquake loading is assumed to be independent of the system being loaded. This decoupling of the system and earthquake loading does not reflect the reality and can have severe consequence in both magnitude and frequency content of the system response. In the current approach the earthquake load is assumed as either the ground motion recorded at ground surface free of any influence of adjacent structures or is simulated based on a target design spectrum. An aspect of the complexity of the system is, as far as structures are concerned, that they are a closed system, while in the case of the supporting soil the system is without boundary. Consequently, the soil-foundation-structure systems are not unique with respect to their properties because of the strain-dependent nature of their characteristics.

In reality, structures are not isolated, especially in intensely populated regions. Adjacent structures influence each other while responding to earthquakes. This influence results from the interaction between each structure with the supporting soil, between adjacent structures through common supporting ground and the alteration of incoming seismic waves from the summation of the interaction of all clustered structures. In addition, the compounding effect of high strain cyclic response of cohesionless soil will induce highly nonlinear behaviour leading to a strong local site effect. This strong nonlinear site response can have severe consequence for structures. This aspect cannot be considered in the design of structures when a fixed base is assumed.

This workshop is one of the activities of the research project under the auspices of the Natural Hazards Research Platform entitled "Impact of liquefiable soil on the behaviour of coupled soil-foundation-structure systems in strong earthquakes" funded by the Ministry of Business, Innovation and Employment. The papers and presentations were by invitation only. However, the workshop itself was open to interested members of the public. The speakers were experts in their field and originated from Japan, Australia, USA, China, New Zealand, Germany and Chile.

The proceedings consisted of 19 papers, which were revised by the authors after the workshop. The editors consider that a revision of the papers will lead to enhanced understanding of the seismic performance of multiple coupled systems. This understanding when incorporated in everyday engineering design has the potential to facilitate a safer seismic environment at a lower cost, especially in large cities of the world.

Nawawi Chouw, Rolando P. Orense & Tam Larkin
The University of Auckland, New Zealand

Acknowledgements

The organizing committee would like to express gratitude for the financial support provided by the Ministry of Business, Innovation and Employment (MBIE) through the Natural Hazards Research Platform (NHRP) and by the Earthquake Commission (EQC). The committee would also like to thank the staff and students of the Department of Civil and Environmental Engineering, the University of Auckland, for making the workshop a success. Many colleagues travelled great distances to attend and present in the workshop. Sincere thanks to all colleagues and friends, especially those who contributed papers to the workshop and actively participated in the discussions. The committee is also appreciative of the review panel who took the time to provide constructive comments to the authors.

Seismic Performance of Soil-Foundation-Structure Systems – Chouw, Orense & Larkin (Eds)
© 2017 Taylor & Francis Group, London, ISBN 978-1-138-06251-1

Organising committee

Prof. Michael J. Pender, *University of Auckland, New Zealand*
Assoc Prof. Nawawi Chouw, *University of Auckland, New Zealand*
Assoc Prof. Rolando P. Orense, *University of Auckland, New Zealand*
Dr Tam Larkin, *University of Auckland, New Zealand*

Seismic Performance of Soil-Foundation-Structure Systems – Chouw, Orense & Larkin (Eds)
© 2017 Taylor & Francis Group, London, ISBN 978-1-138-06251-1

Panel of reviewers

Each paper included in this volume has been carefully reviewed for relevance to the workshop theme as well as for quality of technical content and presentation by at least two members of a panel consisting of the following experts:

Alexei Murashev
Andrew Chan
Chern Kun
Ellys Lim
Esteban Saez
Gonzalo Barrios
Hinke Osinga
Hong Hao
Horst Werkle
Ikuo Towhata
Ilias Dimitrakopoulos

Izuru Takewaki
Jeremy Toh
Luke Storie
Michael Pender
Nawawi Chouw
Rolando Orense
Susumu Iai
Tam Larkin
Xiaoyang Qin
Yukio Tamari

List of participants

Name	Affiliation	Email
Barlow, John	CCL 2015 Ltd	john@coco.co.nz
Barrios, Gonzalo	University of Auckland	gbar737@aucklanduni.ac.nz
Bray, Jonathan	University of California, Berkeley	jonbray@berkeley.edu
Chan, Andrew	University of Tasmania	Andrew.Chan@utas.edu.au
Chan, Yan	KGA Geotechnical Ltd	yan@kga.co.nz
Chin, C.Y.	BECA Auckland	cy.chin@aecom.com
Chouw, Nawawi	University of Auckland	n.chouw@auckland.ac.nz
Cook, Philip	CCL 2015 Ltd	phil@coco.co.nz
Dimitrakopoulos, I.	HKUST	ilias@ust.hk
Grimshaw, Lily	Gaia Engineers Ltd	lily.grimshaw@gaia-engineers.co.nz
Hao, Hong	Curtin University	hong.hao@curtin.edu.au
Iai, Susumu	Kyoto University	iai@geotech.dpri.kyoto-u.ac.jp
Kun, Chern	University of Auckland	ckun005@aucklanduni.ac.nz
Larkin, Tam	University of Auckland	t.larkin@auckland.ac.nz
Leadbeater, J.	Aurecon Christchurch	J.Leadbeater@aurecongroup.com
Lo, Jennifer	KGA Geotechnical Ltd	Jennifer.Lo@kga.co.nz
Loo, Wei Yuen	Unitec	wloo002@aucklanduni.ac.nz
Lim, Ellys	University of Auckland	elim882@aucklanduni.ac.nz
Ma, Martin	Gaia Engineers Ltd	martin.ma@gaia-engineers.co.nz
Mc Curthy, Terry	Soil+Rock Consultants	terry@soilandrock.co.nz
Millen, Maxim	University of Canterbury	mmi46@uclive.ac.nz
Murashev, A. K.	OPUS	alexei.murashev@opus.co.nz
Orense, Rolando	University of Auckland	r.orense@auckland.ac.nz
Osinga, Hinke M.	University of Auckland	H.M.Osinga@auckland.ac.nz
Pang, Chong Heng	Bloxam Burnett & Olliver Ltd	pang@bbo.co.nz
Pender, Michael	University of Auckland	m.pender@auckland.ac.nz
Philpot, Johnny	CCL 2015 Ltd	johnnyp@coco.co.nz
Qin, Xiaoyang	University of Auckland	xqin009@aucklanduni.ac.nz
Saez, Esteban	Pontificia Uni Catolica de Chile	esaez@ing.puc.cl
Sandquist, Rob	Babbage Consultants Ltd	rob.sandquist@babbage.co.nz
Sit, Julia	Auckland	juliasit@hotmail.co.uk
Stapleton, Melcolm	Babbage Consultants Ltd	mjds@babbage.co.nz
Stokoe, Kenneth	University of Texas	k.stokoe@mail.utexas.edu
Storie, Luke B.	Tonkin and Taylor	LStorie@tonkintaylor.co.nz
Tamari, Yukio	Tokyo Electric Power Services Ltd	etamari@tepsco.co.jp
Toh, Jeremy	PSM Australia	jcwtoh@hotmail.com
Towhata, Ikuo	University of Tokyo	towhata.ikuo.ikuo@gmail.com
Werkle, Horst	University of Konstanz	Horst.Werkle@htwg-konstanz.de
Wu, Calvin	Engineering Geology Ltd	calvin.wu@enggeo.co.nz
Yan, Ryan	University of Auckland	r.yan@auckland.ac.nz
Yim, Raymond	Soil+Rock Consultants	raymond@soilandrock.co.nz

Seismic Performance of Soil-Foundation-Structure Systems – Chouw, Orense & Larkin (Eds)
© 2017 Taylor & Francis Group, London, ISBN 978-1-138-06251-1

Photos

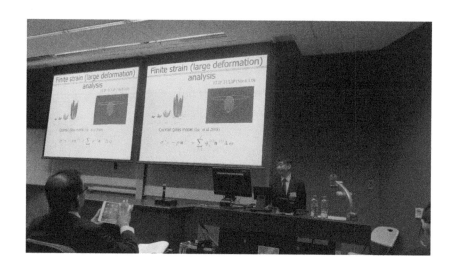

Seismic Performance of Soil-Foundation-Structure Systems – Chouw, Orense & Larkin (Eds)
© 2017 Taylor & Francis Group, London, ISBN 978-1-138-06251-1

Simulation of subsea seismic ground motions

H. Hao & K. Bi
School of Civil and Mechanical Engineering, Centre for Infrastructure Monitoring and Protection, Curtin University, Bentley WA, Australia

C. Li
School of Civil Engineering, Zhengzhou University, Zhengzhou, China

H. Li
Faculty of Infrastructure Engineering, Dalian University of Technology, Dalian, China

ABSTRACT: Seismic motions at seafloor are different from those at onshore sites since seawater can significantly suppress the seafloor vertical motions at the P-wave resonant frequencies of the seawater layer. Seawater layer also indirectly influences the seafloor motions as it changes the water saturation level and pore pressure of subsea soil layers. This paper summarizes some recent developments on the study of characteristics of subsea ground motions, and simulations of seismic motions at seafloor for use in seismic response analysis of offshore structures. The theoretical seafloor transfer functions are presented first. Procedures of simulating seafloor motions from the base rock motion spectrum or using onshore earthquake recordings are introduced. A numerical example is provided following the corresponding theoretical background in each section to demonstrate the procedure. The procedures can be used to simulate seafloor motions that can be used as inputs in the seismic response analysis of offshore structures.

1 INTRODUCTION

In the past few decades, more and more offshore structures such as offshore bridges, pipelines and wind turbines have been constructed. Many of these offshore structures are located in the seismic prone areas. In seismic response analyses of offshore structures, the recorded or synthesized onshore seismic motions are commonly used as the inputs due to the scarcity of offshore seismic recordings and the lack of technique in simulation of seafloor seismic motions. This application may result in inaccurate estimations of offshore structural seismic responses since the characteristics of onshore and offshore seismic motions are different.

Recently, some researchers (e.g. Diao et al. 2014; Chen et al. 2015) studied the characteristics of offshore seismic motions based on the data recorded by the Seafloor Earthquake Measurement System (SEMS) deployed off the coast of Southern California and the strong-motion seismography network (K-net) located throughout Japan. The analyses revealed that the vertical component of seafloor motions is much smaller than that of the onshore motions.

Some theoretical analyses (e.g. Crouse and Quilter 1991; Boore and Smith 1999) were also carried out to investigate the influence of seawater layer on the offshore site ground motion transfer function. It was found that the vertical component of offshore motions can be significantly suppressed by the seawater at the P-wave resonant frequencies of the water layer. Moreover, some other researchers (e.g. Yang and Sato 2000; Wang and Hao 2002) found that seawater layer can indirectly influence the seafloor motions by increasing the degree of saturation and pore pressure of subsea sediments. These two effects of seawater layer can significantly influence the seismic P-wave propagation in porous soil layers and the site amplification of seafloor vertical motion. More recently, Li et al. (2015) theoretically derived the offshore site ground motion transfer functions based on the fundamental hydrodynamics equations and one-dimensional wave propagation theory. This model is believed capable of yielding more realistic estimations of offshore ground motion transfer functions compared to the previous models since the direct and indirect influences of seawater layer can be conveniently considered and the effects of incident angle and the contribution of SV-wave can be considered in the theoretically derived transfer function model. This theoretically derived transfer function model for ground motion at seafloor is briefly introduced in Section 2 of this paper.

After the offshore site ground motion transfer functions are derived, the seafloor seismic motions can be simulated. Some recent works carried out by the authors in this field are summarized in the present study. Particularly, Section 3 introduces the simulation of seafloor motions at a single location and multiple locations (spatially varying seafloor motions) based on the assumed base rock motions, and Section 4 presents the simulation of seafloor motions using onshore earthquake recordings. A numerical example is presented in each section to demonstrate the corresponding method.

2 OFFSHORE SITE GROUND MOTION TRANSFER FUNCTIONS

2.1 Theoretical background

Figure 1a shows a typical onshore site with multiple soil layers resting on a half space (base rock). The base rock properties are characterized by the shear modulus (G_B), damping ratio (ξ_B), density (ρ_B) and Poisson's ratio (v_B). For the multiple soil layers (N layers of soil are shown in the figure), the corresponding parameters are soil thickness (d_i), shear modulus (G_i), damping ratio (ξ_i), density (ρ_i), Poisson's ratio of the soil skeleton (v_i'), soil porosity (n_i) and water saturation degree (S_{ri}), where i represents the ith soil layer. For such an onshore site, the one-dimensional (1D) wave propagation theory proposed by Wolf (1985) was commonly used to predict the local site transfer functions.

In Wolf's theory, the base rock motions are assumed to consist of out-of-plane (y direction in Figure 1a) SH-wave and combined in-plane (x and z directions) P- and SV-waves with respective incident angles. The dynamic equilibrium equation of the site can be expressed in the frequency domain as (Wolf 1985):

$$[S_{SH}]\{u_{SH}\} = \{P_{SH}\} \text{ and } [S_{P-SV}]\{u_{P-SV}\} = \{P_{P-SV}\} \tag{1}$$

where $\{u_{SH}\}$ and $\{u_{P-SV}\}$ are the out-of-plane and in-plane displacements corresponding to the out-of-plane SH-wave and combined in-plane P- and SV-waves. $\{P_{SH}\}$ and $\{P_{P-SV}\}$ are the dynamic load vectors corresponding to the motions in different directions. $[S_{SH}]$ and $[S_{P-SV}]$ are the total stiffness matrices in the out-of-plane and in-plane directions, respectively. They can be formed by assembling the dynamic stiffness matrix of each soil layer in the

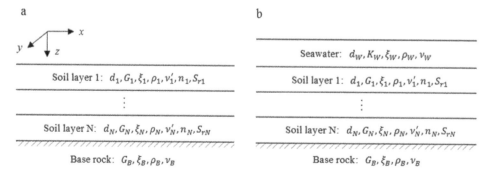

Figure 1. A typical layered (a) onshore site and (b) offshore site.

corresponding direction with the base rock stiffness matrix. By solving Equation 1 in the frequency domain at every discrete frequency, the relationship of the amplitudes between the base rock and each soil layer can be formed, and the site transfer functions in the three different directions ($H_x(i\omega)$, $H_y(i\omega)$ and $H_z(i\omega)$) at each soil layer can be formulated.

It should be noted that the influence of water saturation was not considered in Wolf's theory. To take into consideration the water saturation effect on site amplification, the porosity n, the degree of saturation S_r and bulk modulus of homogeneous fluid K_f proposed by Yang and Sato (2000) can be introduced. The revised Poisson's ratio and P-wave velocity of porous soil can be formulated and substituted into Equation 1 to obtain the site transfer functions (Yang and Sato 2000; Wang and Hao 2002).

Figure 1b shows a typical offshore site. Compared to Figure 1a, one more layer of seawater is considered. In which K_W is the bulk modulus of seawater. The meanings of other parameters are the same as the soil layers. The transfer functions of this offshore site were theoretically derived by Li et al. (2015), and it is briefly introduced here. More detailed information can be found in Li et al. (2015).

In the derivation, the seawater was regarded as an ideal fluid and can only propagate seismic P-wave. The dynamic stiffness of seawater layer was derived based on the fundamental hydrodynamic equations for ideal fluid, i.e. the conservation equation of mass, the Euler's equation and the adiabatic equation of state (Versteeg and Malalasekera 2007). After some lengthy derivations, the dynamic stiffness matrix of the seawater layer was formulated as (Li et al. 2015):

$$[S_P^W] = \frac{K_W^* \omega}{l_x s c_p^* \sin(ksd_W)} \begin{bmatrix} \cos(ksd_W) & -1 \\ -1 & \cos(ksd_W) \end{bmatrix} \tag{2}$$

where K_W^* is the complex value of the fluid bulk modulus, ω is frequency, l_x is the propagation direction cosine in the Cartesian coordinate system, $s = -i\sqrt{1 - 1/l_x^2}$, c_p^* is the complex P-wave velocity and k is wave number. It can be seen from Equation 2 that $[S_P^W]$ is a symmetric matrix and it can be determined by the seismic incident angle and the fluid property.

By assembling the derived dynamic stiffness matrix of the seawater layer with those of the soil layers and base rock, $[S_{P\text{-}SV}]$ in Equation 1 can be obtained. Similar to the derivation of onshore site transfer function as mentioned above, the in-plane transfer functions of offshore site can be calculated by solving Equation 1. Since $[S_{P\text{-}SV}]$ consists of the contribution of seawater, the influence of seawater layer can be conveniently considered. The accuracy of the theoretically derived site transfer function was validated by Li et al. (2015) by comparing the results with the models proposed by Crouse and Quilter (1991) and Boore and Smith (1999), which are special cases of the model by Li et al. (2015), as well as some recorded subsea ground motions. It should be noted that this method can conveniently consider the effect of incident angle and contribution of SV-wave. It also should be noted that since S-wave cannot transmit seawater, $[S_{SH}]$ in Equation 1 is therefore not influenced by the seawater. In other words, the out-of-plane transfer functions $H_y(i\omega)$ for the onshore and offshore sites are the same.

2.2 Numerical example

Figure 2b shows an example offshore site. The parameters of the base rock, soil layers and water layer are taken from the offshore site beneath the SEME station S4IR by Boore and Smith (1999). This site will be used in the following sections as well. For comparison, an onshore site (Figure 2a) is also considered and the onshore site soil properties are revised based on the offshore site data. The meanings of different parameters are defined in Section 2.1. It should be noted that the Poisson's ratio of soil skeleton (v) for both sites are assumed to be the same, but the actual Poisson's ratio of each porous soil layer depends not only on the soil skeleton Poisson's ratio but also on the water saturation level. Poisson's ratio of porous soil increases with water saturation level (Yang and Sato 2000; Wang and Hao 2002). The incident angles of in-plane P-wave and out-of-plane SH wave are assumed to be 60°.

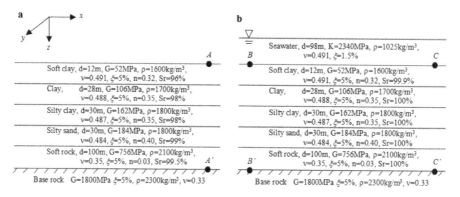

Figure 2. Example onshore and offshore sites, (a) onshore site and (b) offshore site.

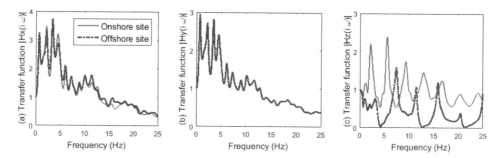

Figure 3. Three-dimensional onshore and offshore site transfer functions.

Figure 3 shows the onshore and offshore site transfer functions for the horizontal in-plane, horizontal out-of-plane and vertical in-plane motions. It can be seen that the transfer functions of horizontal out-of-plane motions for the onshore and offshore sites are the same (Figure 3b). This is because seawater is regarded as ideal fluid and can only propagate seismic P-wave as mentioned above. Therefore it will not influence the propagation of SH-wave. Figure 3a shows the site transfer functions in the horizontal in-plane direction. Seawater layer only slightly changes the site transfer function as shown. However, seawater layer can significantly suppress the vertical in-plane motions especially near the P-wave resonant frequencies of the seawater layer. The resonant frequencies of P-wave in the seawater can be estimated as

$$f_N = N c_p^{*W} / 4 d \sin \beta_p^W \quad (N = 1, 3, 5 \cdots) \tag{3}$$

where $c_p^{*W} = \sqrt{K^W / \rho^W}$ is the P-wave velocity in the seawater, which is 1501.9 m/s based on the water property shown in Figure 2; β_p^W is the P-wave incident angle, which is related to the base rock P-wave incident angle (60° as assumed) and the direction cosine, and it is calculated to be 62.3°. The first three resonant frequencies are therefore 4.35, 13.05 and 21.75 Hz with the water depth of 98 m. Figure 3c clearly shows that the values of offshore site transfer function are generally smaller than those of the onshore site and they are almost zero at these three frequencies.

3 SIMULATION OF SEAFLOOR SEISMIC MOTIONS BASED ON THE ASSUMED BASE ROCK MOTIONS

3.1 Theoretical background

A zero mean stationary stochastic process $x(t)$ can be simulated by the following formula if the power spectral density (PSD) function of the process is known

4

$$x(\mathrm{t}) = \sum_{k=1}^{M} \sqrt{4S(\omega_k)\Delta\omega} \cos(\omega_k t + \phi_k) \tag{4}$$

where M is the number associated with the cut-off frequency in the simulation, $S(\omega)$ is the PSD of the process, ω_k is the kth circular frequency, $\Delta\omega$ is the frequency interval and ϕ_k is a random phase angle uniformly distributed in [0, 2π]. The nonstationary ground motions $a(t)$ can be obtained by multiplying the stationary stochastic process with an envelope function $\zeta(t)$, i.e. $a(t) = \zeta(t)x(t)$.

For an onshore site with multiple soil layers (Figure 1a) or an offshore site with seawater layer (Figure 1b), $S(\omega)$ can be expressed as (Wang and Hao 2002; Bi and Hao 2012; Li et al. 2016):

$$S(\omega) = |\mathrm{H}(\omega)|^2 S_g(\omega) \tag{5}$$

where $\mathrm{H}(\omega)$ is the onshore or offshore site transfer functions in different directions (Figure 3) and $S_g(\omega)$ is the PSDs of base rock motions (locations A', B' and C' in Figure 2). Substituting Equation 5 into Equation 4, the three-dimensional seismic motions at the onshore or offshore site can therefore be simulated.

Many offshore structures such as subsea pipelines are extended along the seabed. Seismic motions at different locations along the structure (e.g. at locations B and C in Figure 2) are different. This is generally known as the ground motion spatial variation effect. Many previous studies revealed that spatially varying ground motions can significantly influence the structural responses. The simulation of spatially varying ground motions have been studied by different researchers (e.g. Hao et al. 1989; Bi and Hao 2012; Li et al. 2016). To simulate the spatially varying seafloor motions, the PSD function $S(\omega)$ in Equation 4 becomes a matrix and has the following form:

$$S(i\omega) = \begin{bmatrix} S_{11}(\omega) & S_{12}(i\omega) & \cdots & S_{1n}(i\omega) \\ S_{21}(i\omega) & S_{22}(\omega) & \cdots & S_{2n}(i\omega) \\ \cdots & \cdots & \cdots & \cdots \\ S_{n1}(i\omega) & S_{n2}(i\omega) & \cdots & S_{nn}(\omega) \end{bmatrix} \tag{6}$$

where n is the number of locations and $S_{jj}(\omega)$ and $S_{jk}(i\omega)$ are the auto and cross PSDs. When local site effect is considered, $S_{jj}(\omega)$ can be calculated by Equation 5 and $S_{jk}(\omega)$ is

$$S_{jk}(i\omega) = H_j(i\omega)H_k^*(i\omega)S_g(\omega)\gamma_{j'k'}(d_{j'k'}, i\omega) \qquad j,k = 1,2,\dots,n \tag{7}$$

where, $H_j(\omega)$ and $H_k(\omega)$ are the site transfer functions at locations j and k, respectively; superscript '*' denotes complex conjugate; $\gamma_{j'k'}(d_{j'k'}, i\omega)$ is the coherency loss function of spatial ground motions at base rock with $d_{j'k'}$ representing the distance between locations j' and k'.

After the cross PSD matrix in Equation 6 is formed, the spatially varying ground motions at different locations along the onshore/offshore site and at the base rock $\left(H_j(i\omega) = H_k(i\omega) = 1\right)$ can be generated based on the spectral representation method (e.g. Hao et al. 1989; Bi and Hao 2012; Li et al. 2016).

3.2 Numerical example

The offshore site shown in Figure 2b is adopted again and the three-dimensional seismic motions at location B are generated. For comparison, the ground motions at location A of the onshore site are also simulated.

The motions on the base rock of the onshore and offshore sites (locations A' and B' in Figure 2) are assumed to be the same, and the horizontal components (the motions in the x and y directions) are modelled by the filtered Tajimi-Kanai PSD function as (Clough and Penzien 1993)

$$S_g(\omega) = \frac{\omega^4}{(\omega_f^2 - \omega^2)^2 + (2\omega_f\omega\xi_f)^2} \frac{1 + 4\xi_g^2\omega_g^2\omega^2}{(\omega_g^2 - \omega^2)^2 + 4\xi_g^2\omega_g^2\omega^2}\Gamma \tag{8}$$

in which ω_g and ξ_g are the central frequency and damping ratio of the Tajimi-Kanai PSD function, ω_f and ξ_f are the central frequency and damping ratio of the high pass filter. The parameters for the horizontal motions are assumed as $\omega_g = 10\pi\,rad/s$, $\xi_g = 0.6$, $\omega_f = 0.5\pi$, $\xi_f = 0.6$ and $\Gamma = 0.00306\,m^2/s^3$. These parameters correspond to a ground motion time history with a duration of $40s$ and peak ground acceleration (PGA) of 0.2 g based on the standard random vibration method (Der Kiureghian 1980). The vertical motion (z direction in Figure 2) on the base rock is also modelled with the same filtered Tajimi-Kanai power spectral density function, but the amplitude is assumed to be 2/3 of the horizontal component.

The base rock wave incident angle is assumed as 60° again as in Section 2.2, so the site transfer functions are the same as those in Figure 3. The sampling frequency and upper cut-off frequency are set to be 100 Hz and $\omega_M = 25\,Hz$ respectively in the simulation. Figure 4 shows the simulated three-dimensional seismic motions at the onshore and offshore sites. As shown, the PGAs of the horizontal motions (a_x and a_y) for the offshore site are close to those of the onshore site due to the similar horizontal transfer functions as shown in Figure 3. For the vertical motion, the PGA for the offshore site is 1.18 m/s², which is much smaller than that of the onshore site (2.58 m/s²), owing to the suppression effect of seawater layer and the influence of water saturation level.

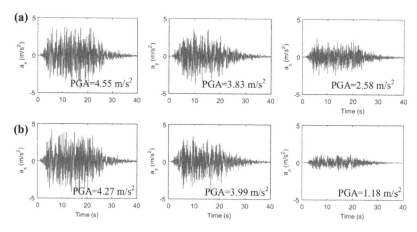

Figure 4. Simulated three-dimensional earthquake motions at (a) onshore site and (b) offshore site.

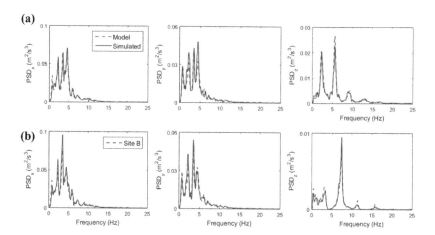

Figure 5. Comparisons of PSD functions of the generated onshore and offshore motions with the corresponding model PSD functions (a) onshore site and (b) offshore site.

Figure 5 shows the comparisons of PSD functions of the generated onshore and offshore motions with the corresponding model PSDs. Good matches are observed. The figures also show that the energy contents are at the same level for the horizontal onshore and offshore motions. For the vertical motion, the energy of the offshore motion is much smaller than that of the onshore site.

To simulate the spatially varying ground motions, the offshore site shown in Figure 2b is adopted again and the three-dimensional earthquake motions at locations B and C are simulated. The Sobczyk model (Sobczyk 1991) is used to describe the coherency loss between the ground motions at points B' and C' on the base rock:

$$\gamma_{B'C'}(i\omega) = \left|\gamma_{B'C'}(i\omega)\right| \exp(-i\omega d_{B'C'} \cos\alpha / v_{app})$$
$$= \exp(-\beta\omega d_{B'C'}^2 / v_{app}) \cdot \exp(-i\omega d_{B'C'} \cos\alpha / v_{app}) \qquad (9)$$

in which $\left|\gamma_{B'C'}\right|$ is the lagged coherency loss, which depicts the similarity between the motions; β is a coefficient which reflects the level of coherency loss, $\beta = 0.005$ is used in the present paper, which represents highly correlated motions; $d_{B'C'}$ is the distance between the points B' and C', and $d_{B'C'} = 200\,m$ is assumed; α is the incident angle of the incoming wave to the site, and is assumed to be 60°; v_{app} is the apparent wave velocity at the base rock, which is 1768 m/s according to the base rock property and the specified incident angle.

Figure 6 shows the simulated acceleration and displacement time histories at locations B and C of the offshore site. Figure 6 clearly shows that the seismic motions at these two locations in a particular direction are different from each other. A time lag caused by the limited apparent wave velocity is clearly shown in Figure 6b.

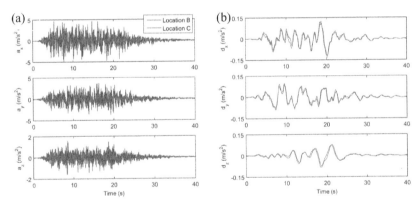

Figure 6. Simulated three-dimensional earthquake motions at locations B and C of the offshore site (a) acceleration time histories and (b) displacement time histories.

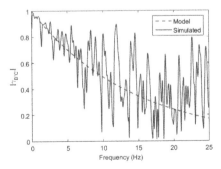

Figure 7. Comparison of coherency loss function between the generated base rock motions with the model coherency loss function.

Figure 7 shows the comparison of the lagged coherency loss function between the simulated base rock motions (between locations B¢ and C¢) with the given Sobczyk model. Good match is observed. It should be noted that for the lagged coherency loss function between the ground surface motions at locations B and C, previous studies (Bi and Hao 2011; Li et al 2016) revealed that it is different from that between the base rock motions, and it is related to the spectral ratios of two local sites. The PSD functions of the generated motions match with the models as well. Due to the page limit, the comparisons are not shown in the paper.

4 SIMULATION OF SEAFLOOR SEISMIC MOTIONS USING ONSHORE EARTHQUAKE RECORDINGS

4.1 *Theoretical background*

This section presents a method to simulate the seafloor seismic motions (e.g. locations E and F in Figure 8) using the more abundant onshore earthquake recordings (at location D). Considering only the linear responses of the sites, the PSDs of the ground motions can be calculated by Equation 5. Particularly, for the onshore site it is

$$S_{on}(\omega) = |H_{on}(\omega)|^2 S_g(\omega) \tag{10}$$

where '*on*' represents the onshore site. When the onshore earthquake recordings are available, the base rock motions can be estimated as $S_g(\omega) = S_{on}(\omega)/|H_{on}(\omega)|^2$. Normally it is reasonable to assume the base rock motions for the onshore and offshore sites are the same, the PSDs for the offshore site motions then become

$$S_{off}(\omega) = |H_{off}(\omega)|^2 S_g(\omega) = |H_{off}(\omega)|^2 /|H_{on}(\omega)|^2 S_{on}(\omega) \tag{11}$$

where '*off*' indicates offshore site. The seismic motions at a single location then can be generated by Equation 4. Similarly, the spatially varying seafloor motions at multiple sites can also be simulated based on the method presented in Section 3.1 when the base rock motions are known.

4.2 *Numerical example*

Figure 8 shows the schematic view of the selected onshore and offshore sites. The soil and water properties for the offshore site are suggested by Boore and Smith (1999) as mentioned above and the onshore site conditions are estimated based on the data recommended by the NGA-West2 Site Data Base (Ancheta et al. 2013). This site is selected because a pair of onshore and offshore seismic motions was recorded by SEME during the 4.8 ML Isla Vista

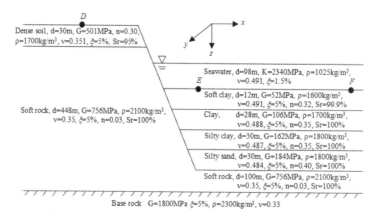

Figure 8.   Schematic view of the layered onshore and offshore sites.

8

Earthquake on 29 May, 2013. The simulated seafloor seismic motions therefore can be compared with the recorded data.

Figures 9 shows the recorded three-dimensional acceleration time histories at the onshore and offshore sites, which are download from the website of US National Centre for Engineering Strong Motion Data (CESMD). It can be seen that the horizontal PGAs at the offshore site are slightly larger than those at the offshore site. However, the vertical PGA of the seafloor motion is much lower as compared with that of the onshore motion due to the influence of water layer as discussed above. Following the method described above, the offshore seismic motions can be generated and Figure 10 shows one realization of the simulated acceleration time histories.

To validate the rationality of the generated seafloor motions, the ratio of vertical to horizontal motions (V/H) is used as an index in the characteristic analysis of seafloor seismic motions since it can remove all the affecting factors on the seismic motions except the effect of local site conditions (Boore and Smith 1999). In the present study the V/H ratios of PGA and response spectrum of the recorded data are compared with the mean values of 800 simulated seafloor motions (it is noted that after 800 simulations the results converge). Table 1 tabulates the results. The averaged V/H response spectral ratio of the simulated seafloor motions (with an assumed damping ratio of 5%) is compared with those of the recorded onshore and offshore motions in Figure 11. It should be noted that the horizontal PGA and response spectrum values are the means of the two horizontal components in the calculation of V/H ratios.

As shwon in Table 1, the mean values of the three-dimensional PGAs and V/H ratios of the generated seafloor motions are close to those of the recorded seafloor motions with a maximum difference of 25.9%. This relatively large difference is expected since only one available

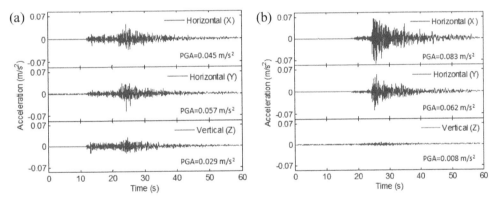

Figure 9. Three-dimensional acceleration time histories recorded by the SEMS stations during the 2013 Isla Vista Earthquake (a) onshore site and (b) offshore site.

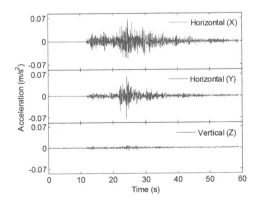

Figure 10. Generated three-dimensional seafloor acceleration time histories.

9

Table 1. PGAs and V/H ratios of the simulated and recorded seafloor seismic motions.

Components	PGA (m/s²)		
	Simulated	Recorded	Difference
Horizontal (X)	0.0815	0.0831	−1.9%
Horizontal (Y)	0.0636	0.0616	3.2%
Vertical (Z)	0.0103	0.0083	24.1%
V/H ratio	0.146	0.116	25.9%

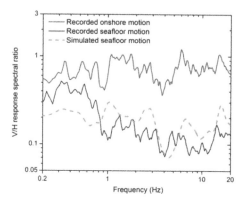

Figure 11. Comparison of the averaged V/H response spectral ratio of the simulated seafloor motions with those of the recorded onshore and offshore motions.

seafloor recording is employed and the uncertainty in the comparison between the simulated and recorded seafloor motions is inevitable. Moreover, the errors in the estimation of local site conditions beneath the onshore and offshore stations will further influence the intensities of the simulated seafloor seismic motions. It can be observed in Figure 11 that the averaged V/H response spectral ratio of the simulated seafloor motions is compatible with that of the recorded seafloor motions. The V/H response spectral ratios of the generated and recorded seafloor motions are much lower than those of the recorded onshore motions. These results are consistent with previous studies. The rationality of the proposed method is therefore validated.

5 CONCLUSIONS

Owing to the lack of recorded subsea seismic ground motion data, most research and design analyses of responses of offshore structures employ onshore ground motions as inputs. Since offshore ground motions have unique characteristics and are different from onshore motions, these practices may lead to inaccurate predictions of offshore structure responses. This paper summaries some of the recent developments in the simulation of offshore seismic motions. The theoretical offshore site transfer functions are presented first, followed by procedures and numerical examples of simulating offshore ground motions based on base rock ground motion spectrum or based on recorded onshore ground motions. These procedures take into consideration the influence of seawater layer on the seismic wave propagations, therefore the simulated ground motions are more representative of subsea motions, and hence lead to better predictions of seismic responses of offshore structures.

ACKNOWLEDGEMENT

The authors would like to acknowledge the partial support from Australian Research Council Discovery Early Career Researcher Award (DE150100195) for carrying out this research.

REFERENCES

Ancheta, T.D., Darragh, R.B., Stewart, J.P., Seyhan, E., Silva, W.J., Chiou, B., Wooddell, K.E., Graves, R.W., Kottke, A.R., Boore, D.M., Kishida, T. & Donahue, J.L. 2013. PEER NGA-West2 Database, *PEER Report No. 2013/03,* University of California, Berkeley.

Bi, K. & Hao, H. 2011. Influence of irregular topography and random soil properties on coherency loss of spatial seismic ground motions. *Earthquake Engineering and Structural Dynamics* 40: 1045–1061.

Bi, K. & Hao, H. 2012. Modelling and simulation of spatially varying earthquake ground motions at sites with varying conditions. *Probabilistic Engineering Mechanics* 29: 92–104.

Boore, D.M. & Smith, C.E. 1999. Analysis of earthquake recordings obtained from the Seafloor Earthquake Measurement System (SEMS) instruments deployed off the coast of southern California. *Bulletin of the Seismological Society of America* 89(1): 260–274.

Chen, B., Wang, D., Li, H., Sun, Z. & Shi, Y. 2015. Characteristics of Earthquake Ground Motion on the Seafloor. *Journal of Earthquake Engineering* 19(6): 874–904.

Clough, R.W. & Penzien, J. 1993. *Dynamics of Structures.* New York: McGraw Hill.

Crouse, C.B. & Quilter, J. 1991. Seismic hazard analysis and development of design spectra for Maul A platform. *Pacific Conference on Earthquake Engineering*, New Zealand.

Der Kiureghian, A. 1980. Structural response to stationary excitation. *Journal of Engineering Mechanics* 106(6): 1195–1213.

Diao, H., Hu, J. & Xie, L. 2014. Effect of seawater on incident plane P and SV waves at ocean bottom and engineering characteristics of offshore ground motion records off the coast of southern California, USA. *Earthquake Engineering and Engineering Vibration* 13(2): 181–194.

Hao, H., Oliveira, C.S. & Penzien, J. 1989. Multiple-station ground motion processing and simulation based on SMART-1 array data. *Nuclear Engineering and Design* 111(3): 293–310.

Li, C., Hao, H., Li, H. & Bi, K. 2015. Theoretical modeling and numerical simulation of seismic motions at seafloor. *Soil Dynamics and Earthquake Engineering* 77: 220–225.

Li, C., Hao, H., Li, H., Bi, K. & Chen, B. 2016. Modeling and simulation spatially correlated ground motions at multiple onshore and offshore sites. *Journal of Earthquake Engineering* 00: 1–25.

Sobczky K. 1991. *Stochastic Wave Propagation.* Kluwer Academic Publishers: Netherlands.

Versteeg, H.K. & Malalasekera, W. 2007. *An introduction to computational fluid dynamics: the finite volume method.* England: Pearson Education.

Wang, S. & Hao, H. 2002. Effects of random variations of soil properties on site amplification of seismic ground motions. *Soil Dynamics and Earthquake Engineering* 22(7): 551–564.

Wolf JP. 1985. *Dynamic Soil-Structure Interaction.* Prentice Hall: Englewood Cliffs.

Yang, J. & Sato, T. 2000. Interpretation of seismic vertical amplification observed at an array site. *Bulletin of the Seismological Society of America* 90(2): 275–285.

Seismic Performance of Soil-Foundation-Structure Systems – Chouw, Orense & Larkin (Eds)
© *2017 Taylor & Francis Group, London, ISBN 978-1-138-06251-1*

Comparative study of deck-abutment interaction with different contact models

Z. Shi & E.G. Dimitrakopoulos
The Hong Kong University of Science and Technology, Hong Kong

ABSTRACT: The earthquake-induced deck-abutment interaction often triggers drastic change of the effective mechanical system of bridge structures. The key to a numerical study of deck-abutment interaction is the simulation of the contact phenomenon. The present paper compares two simulation approaches: a compliance (or gap element) approach and a nonsmooth dynamics approach. The two approaches are assessed with respect to their ability to predict the measured response of a straight, 4-span bridge model in an experimental shake-table study. The paper also investigates the sensitivity of the deck response to the contact element stiffness and the rotation of the abutment. The results show that the deck rotation predicted by these two approaches differs notably.

1 INTRODUCTION

Deck-abutment contact occurs often during earthquakes (Buckle et al. 2012, Kosa et al. 2002). Contact has direct and indirect consequences on the seismic response. The indirect consequences are often more important since contact alters drastically the effective mechanical system, activating behavior unforeseen in the design stage. This discrepancy between the assumed seismic behavior during design and the actual seismic behavior triggered by contact can be detrimental for the bridge, leading even to deck unseating/collapse (Priestley et al. 1996, Yashinsky & Karshenas 2003).

In the heart of any numerical/analytical study of the dynamics of pounding bridge-segments is the simulation of the contact/impact phenomenon at the deck level. Most studies simulate solely the behavior in the normal direction of contact, adopting a gap element ('compliance') approach; i.e. a contact spring (sometimes combined with a dash-pot) working only in compression and activated after the gap-closure. Hence, the deck-abutment interaction is either simulated with a single gap element at each corner of the deck (e.g. in Maragakis & Jennings 1987, Kawashima & Tirasit 2008), or with multiple distributed gap elements aligned perpendicularly along the contact surface in (e.g. in Kaviani et al. 2012, Huo & Zhang 2013). Lastly, Bi & Hao (2013) have proposed a sophisticated three-dimensional finite-element simulation of pounding in segmental straight bridges.

Following a different paradigm, Dimitrakopoulos (2010, 2011) considered the impact between a planar rigid body (the deck) and a rigid half-space (the abutments) as the archetypal mechanical configuration of deck-abutment impact, and examined the problem within the framework of nonsmooth dynamics. Those studies brought forward the role of friction during contact and showed that similar impact-rotation mechanisms characterize also the more complex frictional impacts. Further, Shi & Dimitrakopoulos (2017) extended the nonsmooth dynamics framework to deal with the multi-support excitation, the inelastic behavior of the reinforced concrete piers, and the case of continuous (frictional) contact of a multibody configuration. That study examined the validity and the limitations of the nonsmooth dynamics framework, by simulating the recorded response of a benchmark straight bridge experimental test by Saiidi et al. (2013), and shed light on the physical mechanism behind the rotation of straight bridges taking into account the frictional deck-abutment contact. Unlike the conventional compliance approach which is local in character, the proposed nonsmooth approach is a conceptually straightforward macroscopic (global) simulation: the bridge deck and the abutment seats are assumed as rigid bodies interacting through unilateral contacts (impenetrability constraint). These characteristics are in accordance with the minimal level of input information available for the contact phenomenon in real bridges.

The present study compares the results of the proposed nonsmooth approach (in Shi & Dimitrakopoulos 2017) with those of a conventional compliance simulation. Specifically, the present paper utilizes the experimental data from Saiidi et al. (2013) to assess the validity of the compliance and the nonsmooth simulation approaches. In regards to the compliance method, the study also investigates the sensitivity of the deck response to the stiffness of the contact element and the rotation of the abutment, which is commonly ignored in the practice. The motivation for this study originates from the increased examples of (skew or straight) bridges suffering pounding-induced in-plane deck rotation and the associated need to comprehend the different models in dealing with this problem.

2   BACKGROUND AND OBJECTIVE BRIDGE MODEL

Saiidi et al. (2013) tested experimentally a conventional 4-span reinforced concrete (RC) bridge. The (1/4-scale) bridge model (Figure 1) has a 32.2-m-long, 2.4-m-wide, straight continuous deck supported on 3 bents. The different bent heights result in a slight stiffness eccentricity with respect to the center of mass of the bridge. Independent shake tables control the acceleration in two translational directions at each bent base, while separate actuators control the displacement solely in the longitudinal direction of each abutment seat. The input excitation are 7 sets of ground motion records, with the target peak ground acceleration (PGA) in the longitudinal direction varying from 0.09 g to 1.20 g (Saiidi et al. 2013).

Crucially, Saiidi et al. (2013) reported the substantial in-plane rotation of the deck despite the fact that the bridge was not skew. In particular, during the first 2 excitations, no severe pounding occurs and the recorded rotation is limited. Consequently, the observed rotation is due to the stiffness asymmetry and the slightly different input ground motions among the shake tables. The following 5 excitations, however, result in more severe deck-abutment interactions. The peak response rotation reaches approximately values of 0.01 rad (0.63°) and the residual rotation values of about 0.005 rad (0.20°).

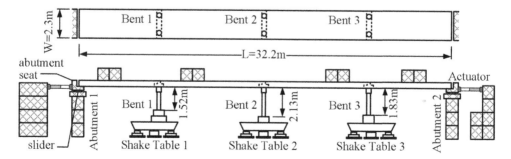

Figure 1.   The 4-span straight bridge configuration tested experimentally in (Saiidi et al. 2013).

3 METHODOLOGY

3.1 *General*

The present study considers the idealized bridge model of Figure 2, subjected to different ground motions (multiple-support excitation) at the 3 bent supports $\left(\mathbf{u}_g^a\right)$ and the 2 abutment seats $\left(\mathbf{u}_a^a\right)$ (Saiidi et al. 2013). The motion of the system is described by 7 degrees of freedom (DOFs) of the three rigid bodies (abutment seat '1' and '2' and the deck 'd', as discussed in Shi & Dimitrakopoulos 2017) with respect to an absolute reference frame (denoted with superscript 'a'). The equation of motion for this multibody system with unilateral contacts and multiple-support excitation can be written as (Dimitrakopoulos 2011, Shi & Dimitrakopoulos 2017):

$$\mathbf{M}\ddot{\mathbf{u}}^a - \mathbf{F}_D\left(\dot{\mathbf{u}}^r\right) - \mathbf{F}_S\left(\mathbf{u}^r\right) - \mathbf{W}_N\boldsymbol{\lambda}_N - \mathbf{W}_T\boldsymbol{\lambda}_T = 0 \tag{1}$$

where $\mathbf{M}$ is the mass matrix; $\ddot{\mathbf{u}}^a$ is the acceleration vector of the deck and the abutment seats with respect to an absolute system of reference (superscript 'a'); $\mathbf{F}_D$ and $\mathbf{F}_S$ are the vectors of the damping and the restoring forces, expressed using the relative velocity/displacement vector $\dot{\mathbf{u}}^r = \dot{\mathbf{u}}^a - \dot{\mathbf{u}}_g^a$ and $\mathbf{u}^r = \mathbf{u}^a - \mathbf{u}_g^a$ respectively; $\mathbf{W}_N$ and $\mathbf{W}_T$ are the direction matrices of the contact (constraint) forces in the normal (subscript 'N') and the tangential (subscript 'T') direction; $\boldsymbol{\lambda}_{NT}$ are the contact force vectors along the two directions of contact. The calculation of the unknown contact force vectors $\boldsymbol{\lambda}_N$ and $\boldsymbol{\lambda}_T$ and direction matrices $\mathbf{W}_N$ and $\mathbf{W}_T$ (Equation 1) depends on the simulation assumptions of the deck-abutment contact/impact interaction.

3.2 *Compliance approach*

The compliance approach is often used in literature in order to calculate pounding forces and the associated local damage (Khatiwada 2014). In particular, most studies model solely the

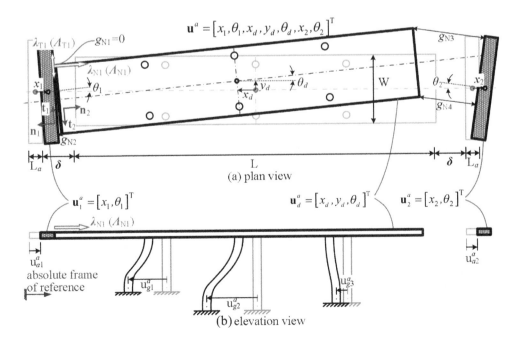

Figure 2. The 4-span straight bridge configuration tested experimentally in (Saiidi et al. 2013).

behavior in the normal direction of contact using the *gap element* (or compliance) approach. The "contact/gap" element is comprised of a spring with, or without, a dashpot to simulate the normal contact force λ_N (Figure 3 (a~b)) (Guo et al. 2015). However, the forces during impact are unpredictable and frail in nature (Khatiwada & Chouw 2014), and hence the contact normal stiffness is still hard to predict reliably (Khatiwada 2014). The following subsections briefly introduce the basic assumptions of the typical gap elements in the longitudinal and the transverse direction, respectively.

3.2.1 *Normal direction*

Typically, the simplest node-to-node contact element is adopted thanks to its straightforward concept and implementation. The gap element is activated when contact occurs, i.e. when the relative distance, i.e. the gap (g_N), between two bodies becomes negative (which implies penetration). Ignoring the rotation of the abutment, the corresponding out normal direction (see Figure 3 (a)) remains unchanged. The gap distance (g'_{N1}) between an upper left corner of the deck and the adjacent abutment seat in this case is (notations explained in Figure 2):

$$g'_{N1} = \delta - x_1 + x_d + 0.5\, L\left(1 - \cos\theta_d\right) - 0.5\, W \sin\theta_d \tag{2}$$

On the contrary, the rotation of the abutment introduces geometric nonlinearity; the normal and the tangential direction of the contact ($\mathbf{W}_N$ and $\mathbf{W}_T$ respectively) and, subsequently, the function of the gap distance (g_{N1}) become response dependent (see Figure 3 (b~c)):

$$\begin{aligned}
g_{N1} &= \cos\theta_1 \left[\delta + 0.5\, L_a - x_1 + x_d + 0.5\, L\left(1 - \cos\theta_d\right) - 0.5\, W \sin\theta_d \right] \\
&\quad - 0.5\, L_a + \sin\theta_1 \left(y_d - 0.5\, L \sin\theta_d + 0.5\, W \cos\theta_d \right)
\end{aligned} \tag{3}$$

The compliance approach calculates the normal contact force (λ_N) based on the penalty method (Wriggers 2006, Guo et al. 2015) either as a linear (e.g. Kelvin-Voigt model with a constant stiffness) or a nonlinear (e.g. Hertz model, Hunt-Crossley model, etc.) function of the penetration g_N. For the sake of the following comparison with the nonsmooth approach, this study adopts the simple Kelvin-Voigt model:

$$\lambda_{Ni} = \begin{cases} 0, & g_{Ni} \geq 0 \\ -k g_{Ni} - c \dot{g}_{Ni}, & g_{Ni} < 0 \end{cases} \tag{4}$$

where i is the index of the contact point, k is the stiffness and c is the damping of the contact element respectively. As a first estimate of the contact stiffness (later in Section 4.2) this study adopts the empirical formula:

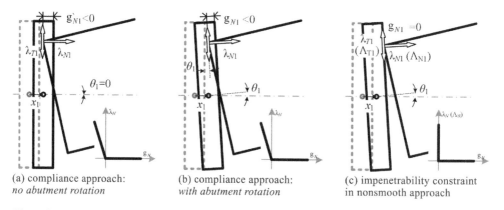

(a) compliance approach: *no abutment rotation*

(b) compliance approach: *with abutment rotation*

(c) impenetrability constraint in nonsmooth approach

Figure 3.   Constraint condition in the normal direction of the contact.

$$k = EA / L \tag{5}$$

where E is the elastic modulus, A is the sectional area, and L is the length of the deck.

3.2.2 *Tangential direction*
The tangential contact force λ_T is obtained in relation to the normal contact force, usually with the regularized Coulomb law among others (Wriggers 2006, in Figure 4):

$$\lambda_{Ti} = \begin{cases} -\mu \text{sgn}(\dot{g}_{Ti})\lambda_{Ni}, & |\dot{g}_{Ti}| \geq \varepsilon \\ -(\dot{g}_{Ti}/\varepsilon)\mu \text{sgn}(\dot{g}_{Ti})\lambda_{Ni}, & |\dot{g}_{Ti}| < \varepsilon \end{cases} \tag{6}$$

where $\dot{g}_{Ti}$ is the contact velocity in the tangential direction of the i-th node, ε is a scalar parameter which determines the sticking-sliding transition, and μ is the coefficient of friction.

3.3 *Nonsmooth dynamics approach*

The proposed nonsmooth approach (Dimitrakopoulos 2010, 2011, 2013, and Shi & Dimitrakopoulos 2017) deploys an event-based analysis framework (Leine et al. 2003, Glocker 2001, Brogliato 2016) which decomposes the dynamic response into continuous motion (without impact) and discontinuous events. Contacts are distinguished into *impacts* (i.e. instantaneous contact events) and *continuous contacts* of finite duration. The present study formulates both types of contacts as pertinent Linear Complementarity Problems (LCPs, Cottle et al. 1992). Further, assume the deck-abutment interaction follows the principles of unilateral contacts (Pfeiffer & Glocker 2004), or, in simple words, that a kinematic constraint ensures the contacting bodies cannot overlap (i.e. no penetration). During impacts (when the gap presents a zero crossing $g_{Ni} = 0$), all non-impulsive forces are considered negligible, wave effects within the body are ignored and the position of the contacting bodies is assumed fixed. Impacts result in instantaneous velocity changes, making the response discontinuous (nonsmooth). Thus, the proposed nonsmooth approach focuses on the impulses transferred employing momentum-impulse principles during impacts (i.e. instantaneous contacts) and adopting Newton's law in the normal, and Coulomb's law in the tangential, direction of the contact. During continuous contacts of finite duration (simultaneously $g_{Ni} = 0$ and $\dot{g}_{Ni} = 0$) contact forces (λ_N and λ_T) are calculated as Lagrange multipliers (Wriggers 2006) to satisfy the impenetrability constraint conditions (Figure 3 (c)). The following sections introduce briefly only the LCP for continuous contact. Refer to Dimitrakopoulos (2010, 2011, 2013) and Shi & Dimitrakopoulos (2017) for the LCP treating impact.

3.3.1 *LCP for continuous contact and detachment*
The forces produced during the continuous contacts enter the equation of motion (Equation 1) through the λ vectors as Lagrange multipliers. Changes of the contact state, e.g.

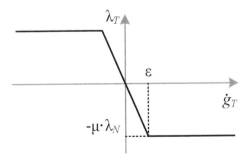

Figure 4.   Regularized Coulomb law in the tangential direction of contact.

17

transitions between sticking and sliding, or between open and closed contacts are captured with the aid of the pertinent contact accelerations or forces. The LCP that treats continuous contacts and captures transitions of the contact states is:

$$\begin{pmatrix} \ddot{\mathbf{g}}_N \\ \ddot{\mathbf{g}}_{HL} \\ \bar{\bar{\boldsymbol{\mu}}}_H \boldsymbol{\lambda}_N + \boldsymbol{\lambda}_H \end{pmatrix} = \begin{pmatrix} \mathbf{G}_{NQ} & -\mathbf{G}_{NH} & \mathbf{0} \\ -\mathbf{G}_{HQ} & \mathbf{G}_{HH} & \mathbf{E} \\ 2\bar{\bar{\boldsymbol{\mu}}}_H & -\mathbf{E} & \mathbf{0} \end{pmatrix} \times \begin{pmatrix} \boldsymbol{\lambda}_N \\ \bar{\bar{\boldsymbol{\mu}}}_H \boldsymbol{\lambda}_N - \boldsymbol{\lambda}_H \\ \ddot{\mathbf{g}}_{HR} \end{pmatrix} + \begin{pmatrix} \mathbf{W}_N^T \mathbf{M}^{-1} \mathbf{h} + \bar{\mathbf{w}}_N \\ \mathbf{W}_H^T \mathbf{M}^{-1} \mathbf{h} - \bar{\mathbf{w}}_H \\ \mathbf{0} \end{pmatrix} \tag{7}$$

$$\begin{pmatrix} \ddot{\mathbf{g}}_N \\ \ddot{\mathbf{g}}_{HL} \\ \bar{\bar{\boldsymbol{\mu}}}_H \boldsymbol{\lambda}_N + \boldsymbol{\lambda}_H \end{pmatrix} \geq \mathbf{0}, \begin{pmatrix} \boldsymbol{\lambda}_N \\ \bar{\bar{\boldsymbol{\mu}}}_H \boldsymbol{\lambda}_N - \boldsymbol{\lambda}_H \\ \ddot{\mathbf{g}}_{HR} \end{pmatrix} \geq \mathbf{0} \text{ and } \begin{pmatrix} \ddot{\mathbf{g}}_N \\ \ddot{\mathbf{g}}_{HL} \\ \bar{\bar{\boldsymbol{\mu}}}_H \boldsymbol{\lambda}_N + \boldsymbol{\lambda}_H \end{pmatrix}^T \begin{pmatrix} \boldsymbol{\lambda}_N \\ \bar{\bar{\boldsymbol{\mu}}}_H \boldsymbol{\lambda}_N - \boldsymbol{\lambda}_H \\ \ddot{\mathbf{g}}_{HR} \end{pmatrix} = 0 \tag{8}$$

with the abbreviations:

$$\mathbf{G}_{NQ} = \mathbf{W}_N^T \mathbf{M}^{-1} \mathbf{W}_Q, \quad \mathbf{G}_{NH} = \mathbf{W}_N^T \mathbf{M}^{-1} \mathbf{W}_H$$
$$\mathbf{G}_{TQ} = \mathbf{W}_T^T \mathbf{M}^{-1} \mathbf{W}_Q, \quad \mathbf{G}_{TH} = \mathbf{W}_T^T \mathbf{M}^{-1} \mathbf{W}_H \tag{9}$$

$$\mathbf{W}_Q = \mathbf{W}_N + \mathbf{W}_G \bar{\bar{\boldsymbol{\mu}}}_G + \mathbf{W}_H \bar{\bar{\boldsymbol{\mu}}}_H \tag{10}$$

where, $\ddot{\mathbf{g}}_N$ is the relative acceleration vector along the normal contact direction and $\boldsymbol{\lambda}_N$ is the normal contact force vector. $\boldsymbol{\lambda}_H$ and $\boldsymbol{\lambda}_Q$ are the tangential contact force vectors of sliding (subscript 'H') and sticking contacts (subscript 'Q'), respectively, $\mathbf{W}_H$ and $\mathbf{W}_Q$ are the corresponding tangential direction matrices for sliding ('H') and sticking contacts ('Q'), respectively. Lastly, $\bar{\bar{\boldsymbol{\mu}}} = \text{diag}\{\mu_i\}$ is a diagonal matrix containing the coefficients of friction.

4 NUMERICAL SIMULATION

4.1 Analytical modelling

Both the compliance and the nonsmooth approaches are implemented on the 4-span straight bridge of Section 2 (Saiidi et al. 2013). Figure 5 presents the global model including the deck, the two abutment seats, and the six columns. There are 4 potential contact points (g_{N1} to g_{N4}). Based on the rigid-body assumption, the motion of the left abutment, the deck, and the right abutment are expressed respectively by $u_1 = [x_1, \theta_1]^T$, $u_d = [x_d, y_d, \theta_d]^T$ and $u_2 = [x_2, \theta_2]^T$. The mechanical configuration of Figure 5 is subjected successively to seven excitations, reproducing the testing procedure followed in the experiments of Saiidi et al. (2013). The nonlinear

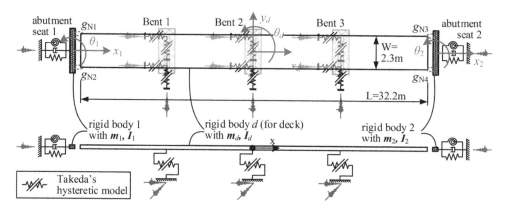

Figure 5.   The 4-span straight bridge configuration tested experimentally in (Saiidi et al. 2013).

18

behavior of the RC columns is modeled with Takeda hysteretic model (Takeda et al. 1970). The calculated contact stiffness based on Equation 5 is $k_0 = 435.8$ kN/mm (equivalent to a unit width stiffness of 381.3 kN/mm/m) which in effect is a numerically "infinite" value. The coefficient of friction in both approaches is $\mu = 0.5$. The remaining detailed parameters of the simulation can be found in Shi & Dimitrakopoulos (2017) but are omitted here for brevity.

4.2 Influence of normal contact stiffness (compliance approach)

The (contact) stiffness of the gap element along the normal direction plays a critical role in the compliance approach. It directly affects the penetration between two bodies and it indirectly affects the frictional force in the tangential direction. Furthermore, the deck-abutment frictional contact force creates a moment about the vertical axis of the deck (Figure 2) which is a dominant cause of the in-plane rotation of the deck. Given the lack of a rigorous procedure to estimate the stiffness of the gap element this section examines values of different orders of magnitude: $10^{+2}k_0$, $10^{+1}k_0$, ..., $10^{-6}k_0$. Note that the analysis in this section considers the abutment rotation, thus the point and direction of contact forces are accounted for properly. In general, the translational response of the deck is similar for all values of stiffness examined hence the following discussion focuses on the penetration during contact and the rotation of the deck.

Figure 6 (top) plots the relative distance of contact point 1 during the first continuous contact event under Excitation No. 7. The contact occurs at the same time instant for all cases, but the lower the stiffness, the larger the penetration and the longer the contact duration. Figure 6 (bottom) compares the response histories of the deck rotation during Excitation No. 7. The peak rotation varies notably according to the assumed contact stiffness. Overall, greater stiffness leads to higher peak rotation. Importantly, not only the peak value but also the response history of the in-plane rotation converge when the stiffness is greater than k_0 (i.e. in the order of magnitude of 10^{+2} kN/mm/m) or smaller than $10^{-4}k_0$ (i.e. in the order of magnitude of 10^{-2} kN/mm/m).

4.3 Influence of abutment rotation

The results of Figure 6 consider the variation of the normal/tangential contact direction introduced by the abutment rotation. However, the most common treatment of the contact problem during design or conventional seismic response analyses is to simply assume a stationary abutment and simulate the motion of the deck by adopting a node-to-node gap element. This section discusses the influence of the abutment rotation on the simulation results. Let the

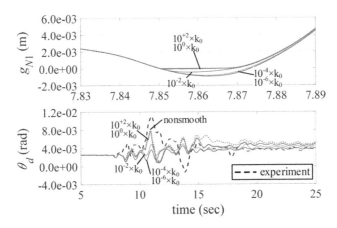

Figure 6. Response histories under Excitation No. 7 of the relative distance of point 1 g_{N1} (top) and the deck rotation θ_d (bottom).

gap function of the former approach be $\mathbf{g}'_N$ (from Equation 2), while of the latter $\mathbf{g}_N$ (from Equation 3), and the pertinent direction matrices $\mathbf{W}_N$ and $\mathbf{W}_T$ are calculated accordingly based on these two equations. Both these two cases of analysis adopt the compliance approach.

Taking into account the rotation of the abutment makes little difference along the two translational directions (the figure is omitted for brevity), but notably affects the response in-plane rotation of the deck. Figure 7 presents two examples of the rotation history during Excitation No. 5 and No. 7. Although the peak rotation in the compliance approach ignoring and considering the abutment rotation shows similar values for Excitation No. 7, the vibration characteristics are significantly different during about 9~13 sec (when contacts are severe, as for example during Excitations No. 5 and 7). Similar differences are also apparent for Excitations No. 3, 4, and 6 but not for Excitations No. 1 and 2 where contact events are few in number and weak in intensity.

Additionally, Figure 8 summarizes the percentage difference for all 3 components of the peak deck response during all 7 excitations. Here, the negative value stands for the case the simulation result taking into account the rotation of the abutment is greater than ignoring it. The comparison shows that ignoring the rotation of the abutment leads to the underestimation of the response of the deck, for all 3 components. This trend is more pronounced for the in-plane deck rotation. For example, during Excitation No. 4 the peak rotation calculated by ignoring the abutment rotation is smaller by about 50% compared to the case takes into account it.

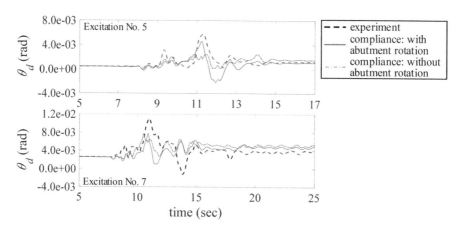

Figure 7. Response histories of the deck rotation under Excitation No. 5 (top) and No. 7 (bottom).

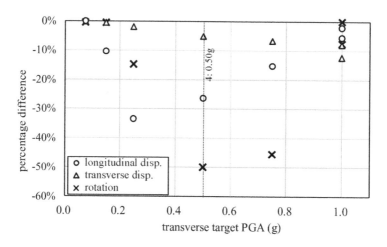

Figure 8. Percentage difference of the peak deck response vs. target transverse PGA.

This section also examines the effect of the geometric nonlinearity introduced by the abutment rotation. Specifically, compared to the case the contact directions (i.e. the direction matrices $\mathbf{W}_N$ and $\mathbf{W}_T$) are response dependent, assuming the contact directions are fixed leads to, a smaller than 1% difference in terms of the peak deck rotation. This negligible difference is attributed to the restricted abutment rotation (maximum about 1.2°) during the shake-table tests. However, further research on the influence of the abutment rotation and the associated change of the contact directions is needed to derive conclusions of general value.

4.4 *Comparison with the nonsmooth approach*

This section compares the results of the compliance and the nonsmooth approaches. In both cases the rotation of the abutment seats is taken into account. The compliance approach simulation assumes: the stiffness of the gap element is $k_0 = 435.8$ kN/mm according to Equation 5; the regularized Coulomb law describes the behavior of contact along its tangential direction with the sticking-sliding transition parameter being $\varepsilon = 0.001$ m/s; and the coefficient of friction is $\mu = 0.5$.

Firstly, consider the results during Excitation No. 5 as an example. The translational displacement in both the longitudinal and the transverse directions (Figure 9) calculated by the compliance approach agrees very well with that of the nonsmooth approach. However, the in-plane rotation of the deck (Figure 9) calculated by the compliance approach exhibits a slightly smaller peak value at about 11.1 sec and from that time point on particular discrepancies appear, when compared to the nonsmooth approach. Besides, the comparison of the contact distance (Figure 10) shows a good agreement for all 4 contact points throughout the analysis.

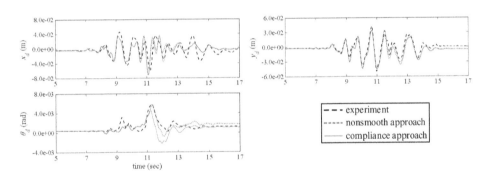

Figure 9.   Comparison of response history of the center of mass of the deck during Excitation No. 5.

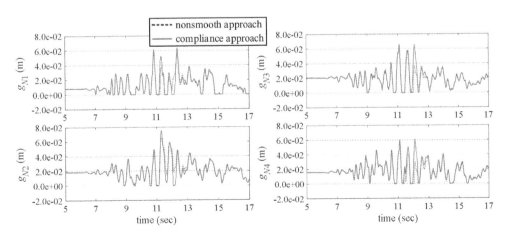

Figure 10.   Contact distance (gap) history during Excitation No. 5.

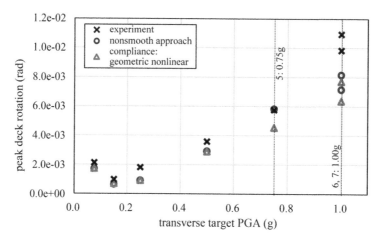

Figure 11. Peak deck rotation vs. transverse target PGA.

Figure 11 summarizes the peak deck rotation calculated by the compliance and the nonsmooth approach with respect to the transverse target PGA. It illustrates that both approaches generally reproduce the trend (observed during the experiments) that the peak deck rotation increases almost exponentially with the PGA of the excitation. Quantitatively, however, the nonsmooth approach yields a closer match although it underestimates the measured peak rotation for the higher intensity Excitations No. 6 and No. 7. On the other hand, the compliance approach leads to the even lower peak deck rotation for Excitation No. 5~7, compared to the nonsmooth approach results.

Recall that the nonsmooth approach calculates the pertinent contact forces (during continuous contacts) not based on an assumed stiffness value, but based on the impenetrability constraint condition. In contrast, the compliance approach calculates the contact forces hinging upon an assumed contact stiffness value. The present compliance simulation uses a practically "infinite" stiffness value which results in almost zero penetration (maximum 1.6 mm). It can be inferred, that this high stiffness resembles the impenetrability boundary condition of the nonsmooth approach and thus leads (along the translational directions) to a similarly good match between the two approaches and the experimental results. Nevertheless, the torsional moment produced by the normal contact forces and the frictional contact forces dominate the in-plane rotation (illustrated in Shi & Dimitrakopoulos 2017). Therefore, even a slight difference of the calculated (normal/tangential) force, or pertinent lever arm, between the two approaches will lead to a significant difference of the deck rotation. Moreover, this difference may accumulate and produce different residual rotations.

5 CONCLUSIONS

The present paper compares a compliance (or gap element) approach with a nonsmooth approach, for the simulation of deck-abutment pounding during seismic response-history analysis of a bridge-abutments system. Specifically, the response of a benchmark straight bridge involving deck-abutment contact, tested experimentally by other researchers, is simulated using the two different methods (compliance approach and nonsmooth approach). The study demonstrates the ability of the nonsmooth approach to satisfactorily reproduce the measured (during the shake-table tests) bridge response. The results also underline the importance of carefully selecting the contact parameters of the compliance approach and the need to take into account the abutment rotation during seismic response analysis. In addition, the paper investigates the influence of the contact stiffness and the rotation of the abutment (i.e. the change of contact normal/tangential direction) on the results of the compliance approach. A parametric study shows that the contact element stiffness and the abutment rotation notably affect the calculated deck response, especially in terms of in-plane rotation.

REFERENCES

Bi K. & Hao H. 2013. Numerical simulation of pounding damage to bridge structures under spatially varying ground motions. *Engineering Structures* 46: 62–76.

Brogliato B. 2016. *Nonsmooth mechanics: models, dynamics and control*. Springer.

Buckle I., Hube M., Chen G., Yen W. & Arias J. 2012. Structural performance of bridges in the Offshore Maule earthquake of 27 February 2010. *Earthquake Spectra* 28(S1): S533–S552.

Cottle R., Pang J. & Stone R.E. 1992: *The linear complementarity problem*. Academic Press.

Dimitrakopoulos E.G. 2010. Analysis of a frictional oblique impact observed in skew bridges. *Nonlinear Dynamics* 60: 575–595.

Dimitrakopoulos E.G. 2011. Seismic response analysis of skew bridges with pounding deck-abutment joints. *Engineering Structures* 33(3): 813–826.

Dimitrakopoulos E.G. 2013. Nonsmooth analysis of the impact between successive skew bridge-segments. *Nonlinear Dynamics* 74: 911–928.

Glocker C. 2001. *Set-Valued Force Laws*. Springer.

Guo A., Cui L., Li S. & Li H. 2015. A phenomenological contact-element model considering slight non-uniform contact for pounding analysis of highway bridges under seismic excitations. *Earthquake Engineering & Structural Dynamics* 44: 1677–1695.

Huo Y. & Zhang J. 2013. Effects of pounding and skewness on seismic responses of typical multi-span highway bridges using the fragility function method. *Jr. of Bridge Engineering-ASCE* 18 (6): 499–515.

Kaviani P., Zareian F. & Taciroglu E. 2012. Seismic behavior of reinforced concrete bridges with skew-angled seat-type abutments. *Engineering Structures* 45: 137–150.

Kawashima K. & Tirasit P. 2008. Effect of nonlinear seismic torsion on the performance of skewed bridge piers. *Jr. of Earthquake Engineering* 12: 980–998.

Khatiwada S. & Chouw N. 2014. Limitations in simulation of building pounding in earthquakes. *International Jr. of Protective Structures* 5(2): 123–150.

Khatiwada S., Chouw N. & Butterworth J.W. 2014. A generic structural pounding model using numerically exact displacement proportional damping. *Engineering Structures* 62–63: 33–41.

Kosa K, Tazaki K & Yamaguchi E. 2002. Mechanism of damage to Shiwei Bridge caused by 1999 Chi-Chi earthquake. *Structural Engineering/Earthquake Engineering, JSCE* 19 (2): 221 s–226 s.

Leine R.I., Van Campen D.H. & Glocker C.H. 2003. Nonlinear dynamics and modeling of various wooden toys with impact and friction. *Jr. of Vibration Control* 9 (1–2): 25–78.

Maragakis E.A. & Jennings P.C. 1987. Analytical models for the rigid body motions of skew bridges. *Earthquake engineering & Structural dynamics* 15: 923–944.

Pfeiffer F. & Glocker C. 2004. *Multibody dynamics with unilateral contacts*. Weinheim: Wiley-VCH.

Priestley, M.N., Seible, F. & Calvi, G.M. 1996. *Seismic design and retrofit of bridges*. John Wiley & Sons.

Saiidi M.S., Vosooghi A. & Nelson R. 2013. Shake-table studies of a four-span reinforced concrete bridge. *Jr. of Structural Engineering, ASCE* 139: 1352–1361.

Shi Z. & Dimitrakopoulos E.G. 2017. Nonsmooth dynamics prediction of measured bridge response involving deck-abutment pounding, *Earthquake Engineering & Structural Dynamics*. DOI: 10.1002/eqe.2863

Takeda T., Sozen M.A. & Nielsen N.N. 1970. Reinforced concrete response to simulated earthquakes. *Jr. of Structural Division, ASCE* 96(12): 2557–2573.

Wriggers P. 2006. *Computational Contact Mechanics*. Springer.

Yashinsky, M. & Karshenas, M.J. 2003. *Fundamental of seismic protection for bridge*. Earthquake Engineering Research Institute Oakland.

Seismic Performance of Soil-Foundation-Structure Systems – Chouw, Orense & Larkin (Eds)
© 2017 Taylor & Francis Group, London, ISBN 978-1-138-06251-1

Numerical simulation of liquefaction effects on adjacent buildings with shallow foundations

R.P. Orense & Y. Hong
Department of Civil and Environmental Engineering, University of Auckland, Auckland, New Zealand

Y. Lu
School of Engineering and Mathematical Sciences, La Trobe University, Victoria, Australia

ABSTRACT: Earthquake-induced liquefaction has caused significant damages to residential houses and commercial buildings following major earthquakes a result of loss in bearing capacity of the foundation ground. While many research studies have focused on the seismic performance of isolated buildings on liquefied ground, very few have investigated the presence of adjacent buildings on the seismic performance of structures on shallow foundations. In this paper, numerical modeling and analyses were performed to examine liquefaction-induced settlements and tilting of buildings with shallow foundations which are adjacent to each other. The effects of the spacing between buildings of certain dimension subjected to shaking were investigated using the finite element effective stress analysis software FLIP. Results indicate that the presence of adjacent structures significantly affects the building performance, both in terms of building settlement/tilt and excess pore water pressure development. Thus, analysing them as isolated cases when nearby buildings are present appears to be a conservative approach.

1 INTRODUCTION

Past severe earthquakes have, time and again, highlighted the impact of soil liquefaction and associated ground deformations to the built environment. Settlement of the foundation ground and the structure constructed on top of it is perhaps one of the most serious effects that engineers should consider. However, due to the constraints and limitations of laboratory-based experiments, most estimations of building settlements induced by soil liquefaction are based on empirical equations or rules-of-thumb that were developed to estimate post-liquefaction consolidation settlement for the free-field conditions, and such free-field condition is surely different from the ones underneath buildings which are investigated in limited scale the literature. Moreover, the geometric/structural properties of the building have influence on the resulting settlements, but the extent of such effects has not yet been well-explained (Dashti et al. 2010). Additionally, other input parameters, such as deviatoric strains resulting from soil-structure interaction (SSI)-induced building ratcheting, input ground motion intensity and building geometry are not taken into account when analysing free-field condition. The development of excess pore water pressure which could induce changes in the ground motion has not been considered directly in the current procedures (e.g. Byrne et al. 2004; Lopez-Caballero & Farahmand-Razavi 2008).

Furthermore, adjacent structures interact during an earthquake through the soil in a phenomenon known as seismic soil-foundation-structure interaction (SSFSI). This interaction is exacerbated with the occurrence of liquefaction in the foundation ground and this affects both the subsoil and structural performance. Figure 1 illustrates case histories of performance of structures adjacent to each other when the foundation ground liquefied. It is apparent that the behaviour of one structure is significantly affected by the presence of the adjacent building.

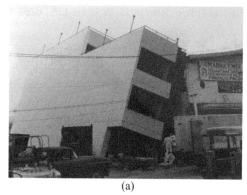 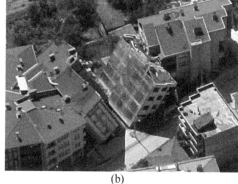

(a)   (b)

Figure 1. Case histories of building performance during earthquakes: (a) 1990 Luzon (Philippines) earthquake (photo by R. Punongbayan); (b) 1999 Kocaeli (Turkey) earthquake (photo from PEER).

Laboratory physical model experiments, such centrifuge tests, have been widely adopted to identify the effects of critical parameters on the soil-foundation-structure interaction, including the inertia forces on the structures and drainage speed (Hausler 2002; Liu & Dobry 1997). For example, Hayden et al. (2015) conducted two centrifuge tests with thorough instrumentation to investigate the response of three types of typical structures founded on liquefiable ground in either isolated or adjacent configurations. For the scenarios examined, they noted that adjacent structures experienced relatively lower foundation accelerations and tended to tilt away from each other, with less settlement compared to the isolated structures. Numerical simulations have also been used to evaluate the basic relationships between the magnitude of building settlement and various factors, such as footing width and thickness of the liquefiable layer (e.g. Dashti & Bray 2013; Shahir & Pak 2010; Hong et al. 2017). However, very limited information about the mechanism of liquefaction-induced building movements has been provided and none of them considered the relative interaction between adjacent buildings.

The main objectives of this paper are: (1) to evaluate the effect of adjacent structures on the degree of settlement and tilt of buildings founded on liquefiable ground through numerical simulation; and (2) to quantify the effects of building spacing on the response of the structures. For this purpose, the effective stress program FLIP was employed.

Before addressing the above objectives, a brief background of the computer program FLIP is discussed, followed by a discussion of the validation of the numerical model adopted.

2 BACKGROUND OF FLIP PROGRAM

There is no doubt that a capable numerical effective stress analysis that is well-calibrated and executed provides the most realistic simulation of the liquefaction process. For this purpose, the computer program FLIP (Finite Element Analysis of Liquefaction Process) was used. The FLIP program was originally developed in Japan by Port and Harbour Research Institute, Ministry of Transport (currently Port and Airport Research Institute) (Iai et al. 1992). The program, especially formulated for dynamic effective stress analysis of soil-structure systems during earthquakes, has been well-validated using Japanese earthquake case histories and has been applied to design many waterfront structures in Japan (FLIP Consortium 2011).

The advanced version of FLIP program has a two-dimensional effective stress analytical scheme that allows redistribution and dissipation of excess pore water pressure based on the constitutive model called *cocktail glass model*. The model assumes that a granular material consists of an assemblage of particles with contacts either newly-forming or disappearing, changing the micromechanical structures during macroscopic deformation. These structures are idealized through a strain space multiple mechanism model as a two-fold structure consisting of a multitude of virtual two-dimensional mechanisms, each of which consists of a multitude of virtual simple shear mechanisms of one-dimensional nature. In particular, a second-order fabric tensor describes direct macroscopic stress–strain relationship, and a fourth-order fabric tensor describes incremental relationship.

In this framework of modeling, the mechanism of interlocking defined as the energy-less component of macroscopic strain provides an appropriate bridge between micromechanical and macroscopic dilative component of dilatancy. Another bridge for contractive component of dilatancy is provided through an obvious hypothesis on micromechanical counterparts being associated with virtual simple shear strain. It is also postulated that the dilatancy along the stress path beyond a line slightly above the phase transformation line is only due to the mechanism of interlocking and increment in dilatancy due to this interlocking eventually vanishing for a large shear strain. These classic postulates form the basis for formulating the dilatancy in the strain space multiple mechanism model. For further details, see the work by Iai et al. (2011).

3 MODEL VERIFICATION OF CENTRIFUGE TEST

Before proceeding with the numerical investigation of the effect of building spacing on their response, the model adopted was first validated through the results of centrifuge tests available in the literature. This is important to ensure that the model adopted generally captures the behaviour of structure in liquefied ground. For this purpose, the centrifuge tests carried out by Dashti et al. (2010) was numerically simulated using the program FLIP. In the process, the required input parameters were established.

3.1 *Determination of soil parameters*

Among the centrifuge tests reported by Dashti et al. (2010), the 1-D test on T3-50-SILT was selected, where the soil profile consists of the following (from the top to bottom): Monterey Sand (relative density, $D_r = 85\%$, 1.2 m thick), Silica flour (0.8 m thick), Nevada Sand ($D_r = 50\%$, 3 m thick) which was liquefiable, and Nevada Sand ($D_r = 90\%$, 21 m thick).

The static properties of the soils, such as density, porosity, shear modulus and bulk modulus, were all determined as reported by Dashti and Bray (2013). On the other hand, the cyclic/dynamic properties are controlled by parameters that are related to the soil behaviour during liquefaction. To obtain the dynamic parameters for input in FLIP, the results of undrained cyclic triaxial tests on the soils need to be numerically simulated. However, cyclic triaxial results for Nevada Sand ($D_r = 50\%$) are not available in the literature; as substitute, Christchurch Sand, which has been extensively investigated at the University of Auckland, was used instead. For comparison purposes, the index properties of both Nevada Sand and Christchurch Sand are summarised in Table 1, while the grain size distribution curves are shown in Figure 2. These properties of Christchurch Sand were determined based on methods specified in NZ Standard (SNZ 1986). Based on these properties, it can be observed that Christchurch Sand is a reasonable substitute to Nevada Sand.

Aside from using Christchurch Sand to calibrate the $D_r = 50\%$ Nevada Sand, the present analysis used the same parameter values as reported by Dashti and Bray (2013) in the numerical simulation using FLIP; the parameters used are summarised in Table 2. In addition to numerically simulating the cyclic resistance curves, other undrained response, such as axial strain development, effective stress path, etc., were matched with the actual experimental results. Figure 3 provides a representative comparison of the simulated and measured

Table 1. Index properties of Nevada Sand and Christchurch Sand.

	Christchurch Sand	Nevada Sand
Specific gravity, G_s	2.65	2.65–2.67
Max. void ratio, e_{max}	1.08	~ 0.75–0.89
Min. void ratio, e_{min}	0.67	~ 0.49–0.56

27

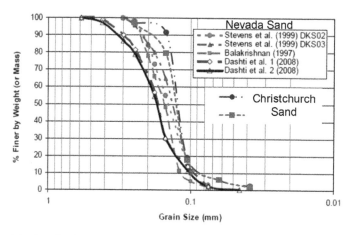

Figure 2. Particle size distribution curves of Nevada Sand and Christchurch Sand (modified from Dashti et al. 2010).

Table 2. Parameters used for the four types of soil layers in the FLIP verification.

Parameters		Nevada Sand ($D_r = 50\%$)	Nevada Sand ($D_r = 90\%$)	Monterey Sand ($D_r = 85\%$)	Silica silt
Density ρ (tonne/m³)		1.63	1.72	1.66	1.67
Porosity n		0.38	0.35	0.36	0.48
Mean eff. conf. pressure (kPa)		100	100	100	100
Initial shear modulus (kPa)		52768	112457	145011	89860
Initial bulk modulus (kPa)		137612	293269	378167	234341
Hydraulic conductivity (m/s)		6×10^{-5}	2.25×10^{-5}	5.29×10^{-4}	3×10^{-8}
Internal friction angle (°)		34	38	40	33
Phase transformation angle (°)		28	29	29	28
Liquefaction parameters for FLIP cocktail glass model[*]	ε_{dcm}	0.2	0.2	0.08	0.2
	r_{edc}	3.6	0.32	0.32	3.6
	r_{ed}	0.25	0.25	0.25	0.25
	r_k	0.73	0.5	0.11	0.5
	q_1	1	1	1	1
	q_2	0.75	1	1	0.75
	q_4	1	1	1	1

[*]For details of these parameters used in FLIP, refer to Iai et al. (1992; 2011).

liquefaction resistance curves and soil response during an undrained cyclic triaxial test on Christchurch Sand (cyclic shear stress ratio, $CSR = 0.18$).

3.2 *Model description*

The configuration of the centrifuge test T3-50-SILT as reported by Dashti et al. (2010) is shown in Figure 4. The thin layer of non-plastic silt (silica flour) was placed on top of the looser layer of Nevada Sand to restrict rapid pore pressure dissipation vertically. In the model, the middle structure represented a 2-story building (height above ground, $H = 5$ m) with width $B = 6$ m; the left structure was wider ($B = 12$ m) and the right one represented a 4-story building ($B = 6$ m, $H = 9$ m). All structural models were treated as rigid elements, which were placed on a 1 m-thick rigid mat foundation. Joint elements are specified at the bottom of the structures to allow sliding and separation between the foundation and the ground. The finite element mesh used in FLIP is illustrated in Figure 5.

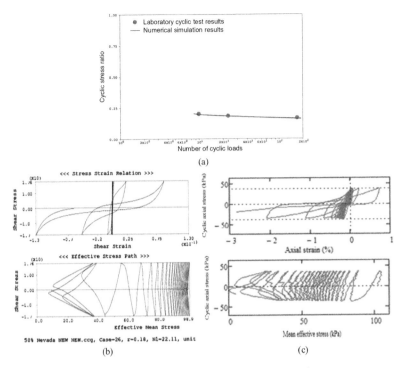

(a)

(b)   (c)

Figure 3. Comparison between: (a) liquefaction resistance curves; results for Christchurch Sand (D_r = 50%) at CSR = 0.18 for (b) simulation and (c) laboratory cyclic triaxial tests.

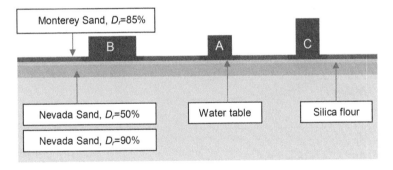

Figure 4. Schematic diagram of the centrifuge model test (modified from Dashti et al. 2010).

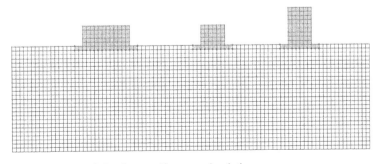

Figure 5. Finite element mesh for the centrifuge test simulation.

3.3 Verification of results

The results of the centrifuge test using the ground motion recorded at a depth of 83 m in the Port Island downhole array during the 1995 Kobe earthquake (with peak base acceleration scaled to 0.55 g) as input motion are compared to the FLIP simulation results in terms of the development of excess pore water pressure ratio and vertical displacements. These are discussed below.

3.3.1 Excess Pore Water Pressure (EPWP) ratio

The excess pore pressure time histories under each structure obtained by FLIP simulation are compared with the results of the T3-50-SILT centrifuge test results in Figure 6. When considering the centrifuge results, refer to those for T3-50-SILT test. It is obvious that the excess pore water pressure generated in the numerical simulation was much higher than in the centrifuge test. The hydraulic conductivity, k, for the silt layer appears to be too low to allow water to be drained out of the ground over a such short period; therefore, high excess pore water pressure is to be expected in the simulation. Note that in the centrifuge test, the 0.8 m thickness of silica silt layer may have cracked due to large ground deformation, and allowed the dissipation of excess pore water pressure; such is difficult to model numerically.

3.3.2 Vertical displacement

Next, the vertical displacements under each structure computed using FLIP were compared with the centrifuge test results, and these are shown in Figure 7. As in the previous discussion, please refer to T3-50-SILT for the centrifuge test results. From the figure, it can be surmised that the residual vertical displacements of the three structures computed from numerical simulation are roughly the same as those from the centrifuge tests. Moreover, the trends of the vertical displacements of the three structures are also very similar to each other.

From the observed results, a couple of observations can be highlighted: (1) when two buildings have the same width, the taller building tends to settle more than the shorter one; and (2) when two buildings have the same height, the narrower building will settle more

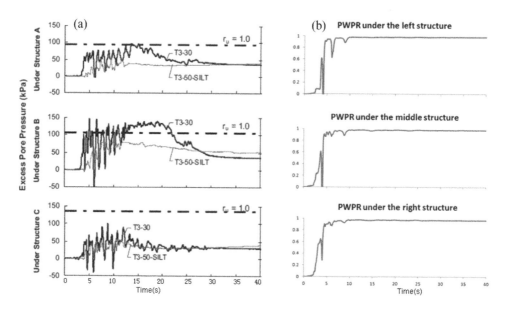

Figure 6.   (a) Centrifuge test results; and (b) simulation results showing the development of excess pore water pressure ratio under each structure (Note: centrifuge test results are from Dashti et al. 2010).

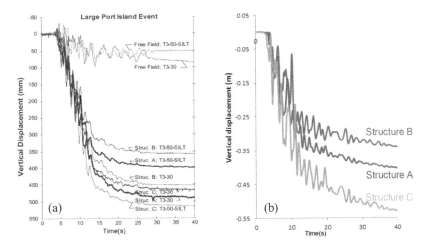

Figure 7. (a) Centrifuge test results; and (b) simulation results showing the development of vertical displacement of each structure (Note: centrifuge test results are from Dashti et al. 2010).

than the wider one. Another observation is that in the relatively "quiet period" after the major shaking (i.e. 15 s to 40 s), the settlements of the three buildings in the centrifuge test became more or less constant, while the numerical simulation showed continuous increase; i.e. the slopes of the vertical displacement curves in the simulation are a bit higher than the experimental ones. A possible reason for this is the quick dissipation of EPWP in the centrifuge test; after $t = 15$ s, the EPWP ratios are almost less than 0.5, indicating that at least half of the effective stress has been restored, resulting in less deformations during the "quiet period".

3.3.3 *Summary of verification exercise*
The numerical simulation of the centrifuge experiment showed satisfactory match, demonstrating that the FLIP effective stress computer program can reasonably simulate the liquefied soil-structure interaction during large earthquakes. The model is then used to investigate the effects of adjacent buildings on the response of buildings constructed on liquefiable ground.

4 NUMERICAL SIMULATION

In the research, the effects of various geometric/earthquake properties on the seismic soil-structure-interaction of building groups were examined. However, due to space limitation, only the effects of the spacing between the buildings are presented, i.e. a building group with three identical structures of specified dimension subjected to a particular acceleration was examined.

4.1 *Input motion*

The input base motion selected was a stochastically generated ground motion; this was adopted so that the characteristics of the motion can be easily controlled in the analysis (Bi & Hao 2012). In FLIP, the horizontal acceleration is applied at the base of the soil profile, which is assumed to be the top of the bedrock. Therefore, the motion selected was generated to fit the shape of spectra for soil site Class A and B (strong rock and rock) as specified in the NZ 1170.5 (SNZ 2004). The input acceleration and its response spectrum are plotted in Figure 8.

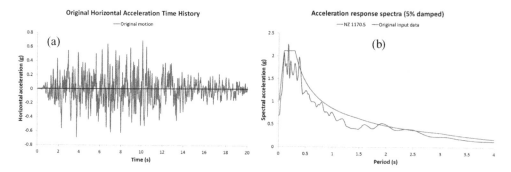

Figure 8. Characteristics of the input acceleration used: (a) time history; and (b) response spectrum at 5% damping and the NZS design spectrum for Class A and B sites.

4.2 Soil-structure model

The model used in the numerical analysis is shown in Figure 9. The model ground consists of three layers: 2 m thick dense crust; 3 m thick liquefiable layer; and 21 m thick dense base layer. The dense materials used in the crust and base layer are the same, i.e. Nevada Sand with D_r = 90%. Similarly, the liquefiable layer is Nevada Sand (with liquefaction properties of Christchurch Sand) with D_r = 50% and represented by cocktail glass model elements with pore water element (drained) for the coupled liquefaction analysis. All the structures considered have similar dimensions: height of 15 m and width of 10 m. They were subjected to the base motion shown in Figure 8 with peak acceleration of 0.5 g. The behaviour was examined considering various spacing between the buildings (i.e. 2 m, 4 m, 6 m, 9 m, 12 m, 15 m). The results are also compared with the case of isolated building of similar dimension. Further details of the numerical modeling, including boundary conditions, are discussed by Hong et al. (2017).

4.3 Simulation results

Figures 10(a) and 10(b) illustrate the vertical displacements and foundation rotation angles, respectively, of the centre building for cases with different spacing between structures. For comparison, the results for isolated building case (called 'reference case') are also plotted in both figures. It is clear that with the increase in spacing between buildings from 2 m to 15 m, the response of the building becomes more similar to that of the isolated structure (reference case). When the spacing between buildings is wider than 9 m, both vertical displacement and foundation rotation angle of the building are similar to the reference case.

When buildings are close enough, such as 2–6 m spacing, the existence of adjacent buildings can be beneficial to each other in resisting the liquefaction-induced building settlement. For example, the total vertical settlement of the building in the case of 2 m gap is only 57% of that in the reference case. When the centre building settles, it displaces the subsoil underneath; but if such soil movement is resisted by the presence of adjacent buildings, then the settlement is mitigated. On the other hand, the maximum and residual (end-of-shaking) foundation rotations of the building increase with reduction in the spacing between buildings; this is because symmetry of induced stresses is lost if adjacent buildings are close to each other.

Figure 11 shows the generation of excess pore water pressure ratio (EPWPR) at the centre of the liquefiable layer directly underneath the centre building. During the first 13 seconds, all curves have almost the same trend. The EPWP remained at a relative high level due to the significant cyclic loading applied induced by the large amplitude of shaking. When the amplitude of shaking started to decrease from t = 12 sec (resulting in decrease in cyclic loading), the EPWP started to dissipate through surface drainage when the building spacing was large; however, for cases when the distance between the buildings was small enough (i.e. 2–6 m), the adjacent buildings provided resistance to water flow inhibiting surface drainage, therefore resulting in such high EPWP to be maintained.

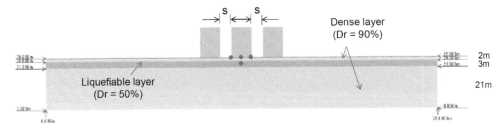

Figure 9. Cross-section of the model ground used in analysis.

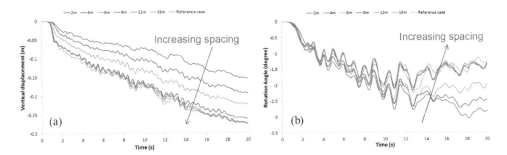

Figure 10. (a): Vertical settlements; and (b) foundation rotation angles for the middle building for different spacing between buildings.

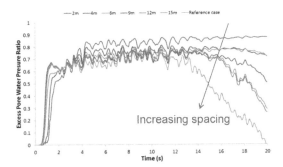

Figure 11. Development of excess pore water pressure ratio at the middle of the liquefiable layer underneath the central building for different building spacing.

To further highlight the above observations, Figure 12 compares the displacement vectors of the soil-structure model and the EPWPR distribution for 2 m gap and 12 m gap at the end of shaking (t = 20 sec). In both cases, non-uniform EPWPR distribution appears to occur underneath the centre structure, with lower (or zero) EPWPR occurring on the left portion which is being compressed as the structure rocks. Larger overall deformation occurs at 2 m spacing, but it is apparent from the displacement vector that the heaving of the soil (for 2 m gap) is restrained by the adjacent building, when compared to the case with 12 m gap. The deformation vectors clearly indicate the effect of spacing between adjacent buildings on subsurface deformations.

Therefore, adjacent buildings can provide three benefits. Firstly, close spacing makes the ground stiffer because the initial elastic stiffness of granular soil is pressure dependent, resulting in reduced strains, displacements and EPWP. Secondly, the existence of the adjacent buildings can resist the occurrence of water drainage (i.e. increasing the drainage path). As mentioned by Dashti et al. (2010), partial drainage combined with the action of cyclic

33

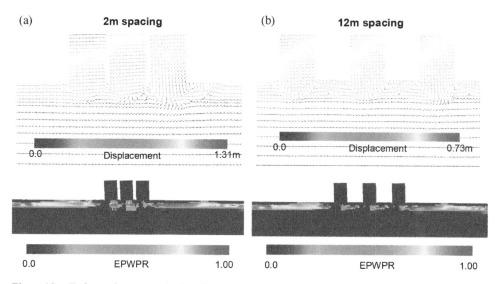

Figure 12.   Deformation vector (top) and excess pore water pressure (bottom) distribution at t = 20 sec ratio for case of: (a) 2 m spacing; and (b) 12 m spacing between buildings.

loading can result in localised volumetric strain; thus, with adjacent buildings, less localized volumetric strain will be induced. Thirdly, for the deviatoric strain, building movements can be the result of the SSI effect. When the building is subjected to cyclic loading, settlement must occur during liquefaction and the surrounding soil will be pushed outward due to structure rocking; however, the existence of adjacent buildings can resist the soil from being pushed away due to the large confining pressure at the surface. Therefore, based on these inferences, buildings in a group in liquefiable ground are expected to perform differently than an isolated structure.

It should be mentioned though that the SSI analyses presented herein are based on standard plane-strain approach for the soil. Under this assumption, the depth of induced stresses by the superstructure is largely overestimated compared to a more realistic 3D case. This tends to increase the SSI, especially when the soil behaves in nonlinear range because a larger portion of surrounding soil is loaded by superstructure. A 3D modelling approach is planned in the future.

5   CONCLUDING REMARKS

Understanding the mechanisms of liquefaction-induced building movements is of great significance to geotechnical engineers in order to mitigate the hazards to buildings caused by large earthquakes. However, there is insufficient research on this subject, especially concerning the mechanisms of building displacements during liquefaction as well as the effect of the presence of adjacent buildings. To provide better insights on these, numerical analyses were performed.

In the verification stage, the use of the finite element analysis software (FLIP) to analyse liquefaction-related issues was validated by simulating a centrifuge experiment and comparing the results with the numerically-derived measurements. Next, the effects of building spacing on the behavior of a building group were examined for a defined building dimension and a specified input ground motion. The results of the analyses indicated that the adjacent buildings, especially when at closer spacing, can resist the movement of the surrounding soil by providing large confining pressure to the surface soil, preventing the displacement of soil below the buildings, as well as higher resistance to drainage. Therefore, from the limited cases investigated, group of buildings in a liquefiable ground is expected to perform differently than isolated ones.

ACKNOWLEDGMENT

The authors would like to thank Prof S. Iai of Kyoto University and the FLIP Consortium for the use of the computer program employed in this research.

REFERENCES

Bi K, Hao H. 2012. Modelling and simulation of spatially varying earthquake ground motions at sites with varying conditions. *Probabilistic Engineering Mechanics*, 29, 92–104.

Byrne PM, Park SS, Beaty M, Sharp M, Gonzalez L, Abdoun T. 2004. Numerical modeling of liquefaction and comparison with centrifuge tests. *Canadian Geotechnical Journal*, 41(2), 193–211.

Dashti S, Bray J, Pestana J, Riemer M, Wilson D. 2010. Mechanisms of seismically induced settlement of buildings with shallow foundations on liquefiable soil. *J. Geotech and Geoenv Eng*, 136(1), 151–164.

Dashti S, Bray JD. 2013. Numerical simulation of building response on liquefiable sand. *Journal of Geotechnical and Geoenvironmental Engineering*, 139(8), 1235–1249.

FLIP Consortium. 2011. *FLIP: Examples of applications*, http://flip.or.jp/e_examples.html.

Hausler EA. 2002. Influence of ground improvement on settlement and liquefaction: A study based on field case history evidence and dynamic geotechnical centrifuge tests. *PhD Thesis*, UC Berkeley.

Hayden CP, Zupan JD, Bray JD, Allmond JD, Kutter BL. 2015. Centrifuge tests of adjacent mat-supported buildings affected by liquefaction. *J. Geotech and Geoenv Eng*, 141 (3), 04014118/1–12.

Hong Y, Lu Y, Orense RP. 2017. Effective stress simulation of liquefaction-induced building settlements. *Proc., 16th World Conference on Earthquake Engineering*, Santiago Chile, 12pp.

Iai S, Matsunaga Y, Kameoka K. 1992. Strain space plasticity model for cyclic mobility. *Soils and Foundations*, 32(2), 1–15.

Iai S, Tobita T, Ozutsumi O, Ueda K. 2011. Dilatancy of granular materials in a strain space multiple mechanism model. *Int. Jour. Numerical and Analytical Methods in Geomechanics*, 35(3), 360–392.

Liu L, Dobry R. 1997. Seismic response of shallow foundation on liquefiable sand. *Journal of Geotechnical and Geoenvironmental Engineering*, 123(6), 557–567.

Lopez-Caballero F, Farahmand-Razavi AM. 2008. Numerical simulation of liquefaction effects on seismic SSI. *Soil Dynamics and Earthquake Engineering*, 28(2), 85–98.

Shahir H, Pak A. 2010. Estimating liquefaction-induced settlement of shallow foundations by numerical approach. *Computers and Geotechnics*, 37(3), 267–279.

Standards New Zealand. 1986. *Methods of Testing Soils for Civil Engineering Purposes*. NZS 4402:1986.

Standards New Zealand. 2004. *Structural Design Actions—Part 5: Earthquake Actions—New Zealand Commentary*. NZS 1170-5 (S1).

Seismic Performance of Soil-Foundation-Structure Systems – Chouw, Orense & Larkin (Eds)
© 2017 Taylor & Francis Group, London, ISBN 978-1-138-06251-1

A study on seismic behavior of foundations of transmission towers during the 2011 off the Pacific coast of Tohoku Earthquake

Y. Tamari, Y. Nakagama, M. Kikuchi, M. Morohashi, T. Kurita, Y. Shingaki, Y. Hirata, R. Ohnogi & H. Nakamura
Tokyo Electric Power Services Co. Ltd., Tokyo, Japan

ABSTRACT: The 2011 off the Pacific coast of Tohoku Earthquake in Japan which occurred 11 March was an earthquake of magnitude of 9.0 with long duration. The electrical accident had been widely occurred in the Kanto region, Japan. The cause of accident was not easily understood, especially concerning the short-circuit accident, which is due to contact of electric wires. To clarify the cause, one of the authors conducted numerical analysis using a model of transmission towers with electric wires (Ohta et al. 2014). Followed by the previous study, we added the conditions regarding foundations and ground on the study which was not considered in the past. First, we investigated ground conditions based on literature, and then conducted soil foundation coupled analysis of two foundations considering different specification of each foundation. It was found that significant difference of earthquake response between foundations which causes contact of electric wire may occur not only for pile-soil systems up to about the depth of 20 m. Other factors such as the difference of shear wave velocity profile of deep part of ground at each foundation may cause occurrence of different earthquake responses.

1 INTRODUCTION

The 2011 off the Pacific coast of Tohoku Earthquake in Japan which occurred 11 March was a massive earthquake of magnitude 9.0 with duration of more than 300 s. The electrical accident had widely occurred in the Kanto region due to the earthquake. The cause of occurrence was not fully understood, especially concerning the short-circuit accident, which is due to contact of transmission wires during the earthquake.

To clarify the cause of the event, one of the authors (Ohta et al., 2014) conducted structural numerical analysis using simulation model with both transmission towers and electric wires considering geometric nonlinearity. They found by the analysis that the contact of wires could occur when phase difference of horizontal displacement between foundations of transmission towers exceeds 0.7 s. But the reason why such amount of phase difference occurred during the earthquake was not fully understood. They mentioned for future study that the accident may possibly occur due to a combination of such various factors as duration of earthquake, structure of towers, earthquake wave, topography, and soil profile around transmission towers.

In this respect, we focused on conditions of not only the structure but also surrounding ground and the pile foundations. It is important to clarify quantitatively the effect of i) foundation and ground condition at subsurface including soil liquefaction, ii) terrain and topography at the site, and iii) the difference of shear wave velocity profile of deep part of ground between foundations.

To start with a series of studies, the authors have conducted three dimensional effective stress dynamic analysis considering soil-structure interaction for two transmission tower foundations in order to examine the occurrence of different earthquake responses between foundations. Since potential of soil liquefaction was reported around the one side of transmission tower by regional government (Yamanashi Prefecture, 2013), the pore water build-up was considered in a part of sandy soil layers. The model used for soil in this study is an extended strain space multiple mechanism model into three dimensional space (Iai, 1993). It is based on a plane strain mechanism (Iai et al., 1992), which has been commonly used for performance based design of various structures in Japan.

2   BRIEF REVIEW OF PAST STRUCTURAL ANALYSIS

The transmission towers are located in the southern part of Kofu basin, Yamanashi prefecture, Japan. Location in detail is illustrated in Figure 1. The event reproduction analysis was conducted using the model composed of 4 towers and 3 span ground and electric wires (Figure 2).

The place of the contact of electric wires which occurred during the 2011 off the Pacific coast of Tohoku Earthquake is marked up in the figure. Length of span between tower No. 2 and No. 3 is 440 m, in which natural period of first mode for electric wire at the span is 7.87 s. Damping ratio of 0.4% for electric wire was considered in the analysis. Program ADINA ver.8.8 (2011) was used for the analysis taking geometrical nonlinearity of electric wires into account.

The observed motion at the recording station "YMN005" (2011/03/11-14:46, 38.103N, 142.860E, M9.0, K-NET: NIED), which was away about 7 km northward from the site, was directly applied to the model. Since it was thought that the phase difference between towers occurred during the earthquake by several causes, variety of phase difference of input motion between foundations from 0.0 s to 5.0 s were considered in the analysis. Results of the analysis were summarized as follows;

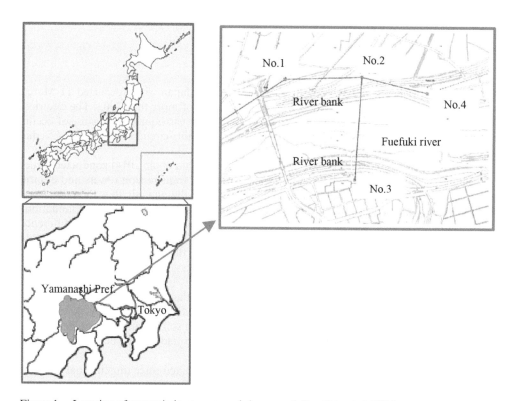

Figure 1.   Location of transmission towers and river area (after Ohta et al. 2014).

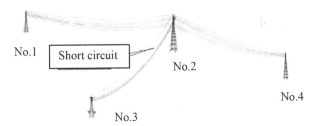

Figure 2.   Numerical analysis model composed of 4 towers and 3 span ground and electric wires (Ohta et al. 2014).

i.   The contact of electric wire which might cause the event of short-circuit was not repro-duced when the same input motion was applied to transmission towers. But it was suc-cessfully reproduced when phase difference of 0.7 to 1.0 s, 3.0 s, 5.0 s were considered.
ii. The place of the contact of electric wire was consistent with the fact evaluation in case the phase difference of 1.0 s was considered. Mode of the motion of electric wire in the verti-cal direction at that time was third or fourth mode rather than the first mode.
iii. The time of contact of electric wire was simulated between 140 s and 200 s, being roughly agreed with the reported result of the event analysis of 120 s to 180 s.

3   DOCUMENT INVESTIGATION ON GROUND CONDITIONS

We examined more details about the conditions of site in geotechnical point of view referring literature.

3.1   *Topography and deep ground velocity structure*

Kofu basin where the site exists exhibits a reverse triangle on the map with the size of about 20 km in east-west direction, 15 km in north-south direction as depicted in Figure 3. The basin is regarded as structural basin which is surrounded by mountains in three directions. The elevation of its bottom surface is about 250 m to 300 m. The site is located at the south tip of lowland of the basin as marked by red circle in the figure. Approximate location of seismic prospecting by Yamanashi regional government (2003) is also depicted in the figure. Figure 4 illustrates geological cross section of Kofu basin along the line of seismic prospect-ing. Evaluated shear wave velocity in the past study at each stratum is written in the legend. Sedimentary stratum and Tertiary formation ($Vs = 1.04$ km/s) exists underlain by top gravel stratum ($Vs = 0.42$ m) with the maximum thickness of about 100 m at the surface.

3.2   *VS30 values of ground*

Figure 5 illustrates VS30 map around the transmission towers (NIED, J-SHIS, 2016). VS30 herein implies average shear wave velocity of surface 30 m subsoil. VS30 value around tower No. 2 is a range of 160 m/s to 200 m/s, whereas the value around tower No. 3 is about 250 m/s to 350 m/s. According to the past evaluation of micro topographic classification, this portion of southern side of the river is regarded as alluvial fan (NIED, J-SHIS, 2016).

3.3   *Liquefaction evaluation*

Yamanashi Regional Government reports result of liquefaction evaluation in whole area of Yamanashi prefecture (Yamanashi Prefecture, 2013). Figure 6 shows distribution of liquefaction potential index (hereinafter called "PL value", Iwasaki et. al. 1981) around the

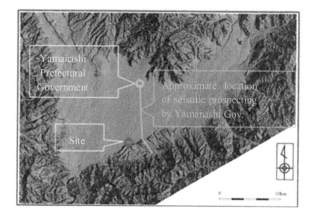

Figure 3.　Relief map of Kofu basin (Yamanashi Pref. Japan, 2003).

Scale of H : Scale of V = 1 : 2

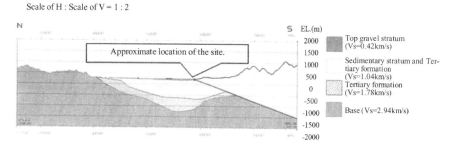

Figure 4.　Cross section of Kofu basin (N-S Direction, Yamanashi Pref. Japan, 2003).

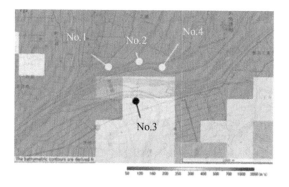

Figure 5.　Map of VS30 value (NIED, J-SHIS, 2016).

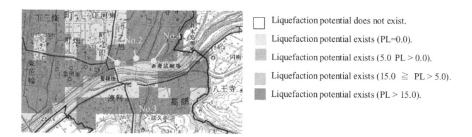

Figure 6.　Liquefaction hazard map around the site (Yamanashi Pref, 2013).

transmission towers. It is seen that PL value around tower No. 2 is more than 15.0 indicating high possibility of liquefaction occurrence. In contrast with this, surrounding ground of tower No. 3 at southern side of river may not liquefy with PL value of 0.0.

3.4 *Consideration*

Taking above mentioned geotechnical conditions into account, it is discussed;

i.   Liquefaction resistance may be lower in subsoil around tower No. 2 than around tower No. 3. This may cause softening of sandy soil during earthquake with long duration, resulting in long natural period of subsoil around tower No. 2.
ii. VS30 value is different at surrounding ground between tower No. 2 and No. 3. Difference of VS30 value is about 100 m/s in the depth of 30 m, causing occurrence of phase difference of earthquake response.
iii. Deep subsurface structure is varying elaborately in the horizontal distance of several hundred meters due to basin terrain, e.g. thickness of gravel stratum (Vs = 420 m/s) is different between the location of tower No. 2 and No. 3. This may cause phase difference of earthquake response.

We consider points of study to clarify the cause of phase difference occurrence in earthquake response from variety of aspects as;

a.   Foundation subsoil system considering liquefaction.
b.   Difference of VS30 value at each transmission tower.
c.   Deep subsoil structure of shear wave velocity and topography of basin.

Point (a) can be considered to conduct soil foundation coupled analysis of transmission tower foundations. (b) and (c) can be taken into account to conduct numerical analysis of large model in which both wide area and large depth of ground are taken into account.

4   SOIL FOUNDATION COUPLED ANALYSIS OF TRANSMISSION TOWER

4.1 *Subsurface soil conditions and model parameters*

Borehole log and N value of standard penetration test (hereinafter called "SPT N value") at each tower are shown in Figure 7. Borehole of tower No. 2 is located at the foundation of tower No. 2, whereas borehole No. 3, which is opened to the public by Ministry of Land, Infrastructure, Transport and Tourism (MLIT, 2016), is about 150 m east from tower No. 3. Test for particle size distribution were conducted using several number of penetration samples. Fine fraction contents of soils are 7.9% to 37.9% in alluvial sand layer, 47.9% to 80.5%

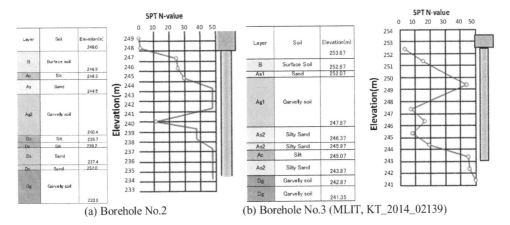

(a) Borehole No.2   (b) Borehole No.3 (MLIT, KT_2014_02139)

Figure 7.   Borehole log and SPT N-value.

in alluvial clayey layer, 5.4% in alluvial gravel layer, respectively. Results of PS logging, cyclic test for dynamic deformation and liquefaction characteristics were not reported.

The ground models at each tower were developed by referring to boring logs. The layer and thickness is illustrated in Table 1 and Table 2. Model parameters of a strain space multiple shear mechanism for soil (Iai et al., 1992, Iai, 1993) were basically assessed using simplified method (Morita et al., 1997) by referring to SPT N value and fine fraction contents. Shear wave velocity were determined using empirical correlation formula with SPT N value (Cabinet Office, Japan. 2001). Parameters for physical characteristics, dynamic deformation characteristics, liquefaction characteristics are summarized in Table 1 and Table 2. Dilatancy of soil was taken into account for sandy bank material below water table. Parameters for liquefaction are specified as best assessed parameter set by numerical simulation of undrained cyclic shear loading.

4.2 Foundations and transmission tower

Each transmission tower has pile foundation with square shaped footing. Presumed approximate depth of footing and pile are illustrated in Figure 7. Specifications of foundations are summarized in Table 3. It is noted that the foundation of tower No. 2 has 8 piles, whereas No. 3 has 12 piles. Footing of foundation is modeled by elastic body, piles by linear beam of which parameters of Young's modulus and poisons ratio are equivalent to those of concrete. Compressive strength of concrete was assumed to be 24 N/mm^2, Young's modulus 25 kN/mm^2, Poisons ratio 0.2, respectively.

The height and natural period of first mode of each transmission tower are shown in Table 4. Tower is simply modeled as single spring-mass system of which natural period is

Table 1. Ground model and parameters for subsoil (Tower No. 2).

Layer	H (m)	ρ (t/m)³	Vs (m/s)	Gma (kPa)	$-\sigma ma'$ (kPa)	ϕf (deg)	$hmax$	ϕp (deg)	Cyc. Str. Ratio DA = 5.0%
B	1.4	1.6	132	27950	13.6	39.0	0.24	–	–
B	0.7	1.6	132	27950	13.6	39.0	0.24	28.0	0.188
Ac	0.6	1.5	161	38760	18.0	30.0	0.20	–	–
As	1.7	1.9	233	103190	24.3	43.9	0.24	–	–
Ag2	1.4	2.0	246	121400	35.2	43.9	0.24	–	–
Ag2	2.8	2.1	266	171700	48.5	43.9	0.24	–	–
Dc	1.2	1.8	271	132140	62.3	0.0[*1]	0.20	–	–
Ds	2.2	1.9	314	186990	71.3	43.8	0.24	–	–
Dg	4.8	2.1	354	263530	95.1	43.9	0.24	–	–

$\Sigma H = 16.8$ m $Vs, ave = 276$ m/s[*1]   Cohesion: 1057 kPa   Ground water level: Ground level–1.4 m (GL–1.4 m).

Table 2. Ground model and parameters for subsoil (Tower No. 3).

Layer	H (m)	ρ (t/m)³	Vs (m/s)	Gma (kPa)	$-\sigma ma'$ (kPa)	ϕf (deg)	$hmax$	ϕp (deg)	Cyc. Str. Ratio DA = 5.0%
B	1.0	1.6	132	27950	13.6	39.0	0.24	–	–
As1	0.6	1.6	131	29200	13.8	38.8	0.24	–	–
Ag1	3.0	1.5	241	116450	36.7	43.9	0.24	–	–
Ag1	1.2	1.9	241	116450	36.7	43.9	0.24	–	–
As2	4.0	2.0	190	68920	75.9	40.0	0.24	–	–
Dg	11.7	2.1	354	262480	129.6	43.8	0.24	–	–

$\Sigma H = 21.5$ m $Vs, ave = 285$ m/s   Ground water level: Ground level–4.6 m (GL–4.6 m).
H: layer thickness; p: density; Vs: shear wave velocity; Gma: elastic shear modulus at a confining pressure of $(-\sigma ma')$; $-\sigma ma'$: reference confining pressure; ϕf: shear resistance angle; and ϕp: phase transformation angle.

Table 3. Specification of pile foundations of transmission tower.

Items	Tower No. 2	Tower No. 3
Dimension of footing	14.8 m × 14.8 m	12.8 m × 12.8 m
Thickness of footing	2.0 m	1.6 m
Types of piles	Cast in Place Pile (RC)	Cast in Place Pile (RC)
Number of pile	8	12
Pile diameter	1200 mm	1200 mm
Pile bottom depth	Ground level–12.50 m	Ground level–9.50 m

Table 4. Natural period of transmission towers.

Items	Tower No. 2	Tower No. 3
Height	74 m	44 m
Natural period of the first mode	0.59 sec	0.29 sec
Damping ratio	0.02	0.02

equivalent to the natural period of first mode of tower. The mass is specified from weight of tower itself excluding the weight of electric wires. The height of mass is simply assumed as one third of total height.

4.3 Finite element model of soil foundation system

Foundation and surrounding ground were modeled in three dimensional finite element model considering detailed shape of footing and piles. The model for tower No. 3 is illustrated in Figure 8 as a typical example. Soil layer was assumed as horizontally layered deposit. Though river bank of several meters high exists near both transmission towers, ground surface is assumed as flat for simplicity. The width and breadth of ground was determined as 90 m, which was more than five times of the foundation width. The elevation of the bottom of model was unified to be EL+232.0 m for both models in order to examine the difference of earthquake response at the same base condition. Side viscous boundary was defined at each side of the model to perform the same seismic behavior as free field at its edge. Bottom viscous boundary was defined as well through which outcrop input motion could be applied from the bottom.

In order to consider the volume of pile (Diameter = 1200 mm) which was modeled by beam element, the cylindrical shape was precisely modeled in finite element mesh of ground. Nodal points in pile beam element were connected by rigid beam to nodal points in soil elements at outer surface of pile. The tower was modeled as single spring-mass model on the footing of foundation.

Damping of model was considered for both soil and foundations by Reyleigh damping of $\alpha = 0.0$, $\beta = 0.001$. The value of β for soil and foundation was determined considering initial natural period of first mode for ground model (No. 2: 0.27 s, No. 3: 0.33 s). It was assumed that initial damping ratio was 0.01. Reyleigh damping for tower of $\beta_{stru} = 0.004$ for No. 2, $\beta_{stru} = 0.002$ for No. 3 was used individually so that it became equivalent to damping of tower (h = 0.02).

4.4 Reproduction of ground motion

The earthquake motion at the base was reproduced for this case study. The observed accelerations at the surface of recording station K-NET Kofu (YMN005, 2011/03/11–14:46, 38.103N, 142.860E, M9.0, NIED) was used for calculation. Location of recording stations was about 7 km north from the site. The outline of ground model used for reproduction calculation of base motion is presented in Figure 9(a), and the peak ground accelerations in Figure 9(b).

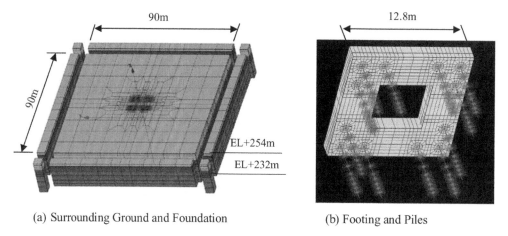

(a) Surrounding Ground and Foundation   (b) Footing and Piles

Figure 8.   Three dimension finite element model (Foundation of tower no. 3).

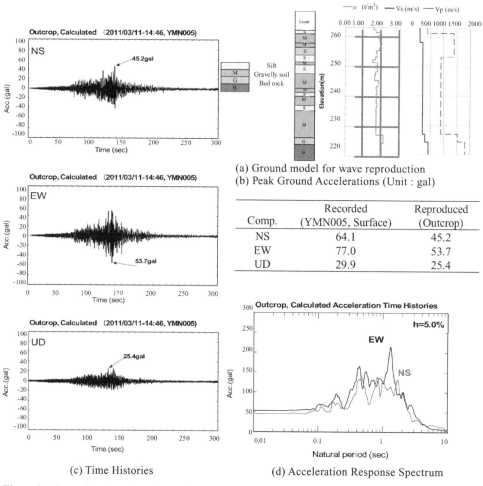

(a) Ground model for wave reproduction
(b) Peak Ground Accelerations (Unit : gal)

Comp.	Recorded (YMN005, Surface)	Reproduced (Outcrop)
NS	64.1	45.2
EW	77.0	53.7
UD	29.9	25.4

(c) Time Histories   (d) Acceleration Response Spectrum

Figure 9.   Reproduced earthquake motion at the base (Vs = 420 m/s).

The model from the surface to ground level–20 m (hereinafter called "GL"–20 m) was developed using borehole data at the recording station by NIED, and the model from GL–20 m to the base of GL–37.7 m (Vs = 420 m/s) by borehole data of deep subsoil structure investigation (Yamanashi Prefecture, 2005), which was located 800 m south from recording station. Degradation of shear modulus and damping ratio with shear strain was specified based on past study (Yamanashi Prefecture, 2005). Calculation was conducted using one dimensional equivalent linear earthquake response analysis based on multiple reflection theory. Reproduced acceleration time histories are illustrated in Figure 9(c), and the response spectrum of the horizontal accelerations in Figure 9(d). Peak base horizontal accelerations resulted in about 50 gal.

4.5   *Earthquake response analysis of soil foundation coupled model*

Before the dynamic analysis, two stages of static analysis were conducted in order to simulate the initial stress of soil and initial section force of piles before the earthquake. First, gravity was applied to ground and foundation, and subsequently, to the transmission tower. Horizontal displacement was constrained at side surface of ground model and both horizontal and horizontal displacement was fixed at the base through static analysis. With these initial condition, an earthquake response analysis was conducted on the soil foundation coupled model. Three components of reproduced motion with duration of 300 s were used simultaneously as the input motion. The analysis was conducted with the undrained conditions (Iai, 1995) in order to simplify the analysis. The time integration was numerically done using the Wilson-θ method (θ = 1.4) using a time step of 0.01 s.

4.6   *Results*

The effective stress analysis results in the maximum decrease of ratio of effective stress $(1 - \sigma_m'/\sigma_{m0}')$ and maximum shear strain γ_{oct}, through the whole durationas illustrated in Figure 10 and Figure 11. It is seen that pore water pressure increased at subsoil around Tower No. 2 up to about ratio of 0.2. The maximum strain is seen to be the order of 10^{-3} in subsoil around tower No. 3 foundation.

Figure 12 illustrates calculated horizontal acceleration time histories at the top of foundation of tower No. 2 and No. 3. Figures (a and b) show time histories for whole duration, and Figures (c and d) for time period when amplitude of acceleration is significant. It is seen that earthquake response between foundations of tower No. 2 and No. 3 are almost the same. Examining the time histories strictly, phase difference of 0.03 s is seen in Figure 12(a) of NS component between the peak of 137.01 s for foundation No. 2 and 137.04 s for foundation No. 3. In reference to the thickness of model and average shear wave velocity shown in

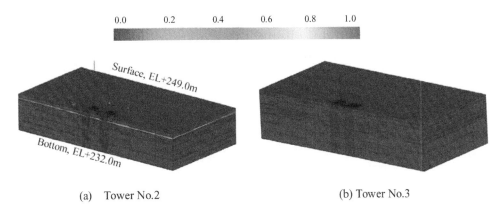

| (a)   Tower No.2 | (b) Tower No.3 |

Figure 10.   Maximum ratio of effective stress decrease ($1-\sigma_m'/\sigma_{m0}'$, 0.0 sec – 300.0 sec).

Table 1 and Table 2, the initial time period of shear wave propagation in model subsoil can be roughly calculated by hand by dividing the thickness by velocity. Those results in 0.061 s (=16.8 m/276 m/s) for No. 2, 0.075 s (=21.5 m/285 m/s) for No. 3 subsoil, being the phase difference of 0.014 s. This phase difference is about half of dynamic analysis of 0.03 s. It is because that the natural period of ground becomes longer due to nonlinear behavior during earthquake, and results in longer period of soil foundation system in both foundations.

The time history of vertical displacements and rotation at the foundation of Tower No. 2 are illustrated in Figure 13. Rotation angle is calculated by dividing the difference of vertical displacement by base distance of tower of 10.959 m. The time history of rotation angle of foundation is seen to have the same trend as the horizontal acceleration time history. However, the maximum value of rotation angle is a relatively small value of about 0.2 to

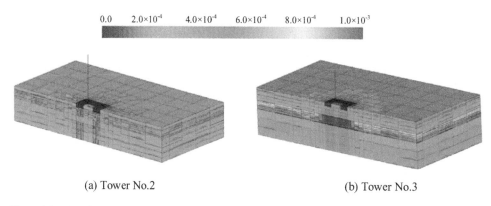

(a) Tower No.2　　　　　　　　　　　　　　(b) Tower No.3

Figure 11.　Maximum shear strain (γ_{oct}, 0.0 sec – 300.0 sec).

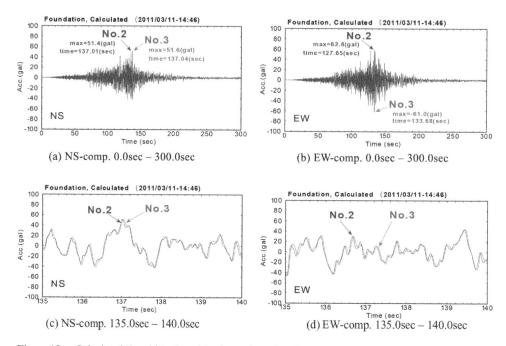

(a) NS-comp. 0.0sec – 300.0sec　　　　　　　(b) EW-comp. 0.0sec – 300.0sec

(c) NS-comp. 135.0sec – 140.0sec　　　　　　(d) EW-comp. 135.0sec – 140.0sec

Figure 12.　Calculated time histories of horizontal acceleration.

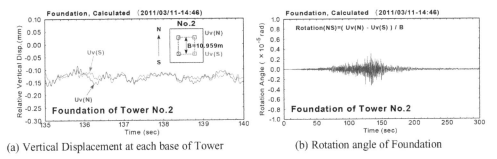

(a) Vertical Displacement at each base of Tower   (b) Rotation angle of Foundation

Figure 13.   Time histories of calculated rotation angle of foundation (Tower No. 2).

0.4×10^{-5} radians. It means that the foundation behaves predominantly by sway motion in the case where the pile tip is placed on hard ground with shear wave velocity of more than 300 m/s.

5   CONCLUSIONS

The present study explained the ground conditions around the transmission towers and the results of dynamic effective stress analysis of tower foundations. The study leads to the following conclusions.

1. Examining the ground conditions based on literature, it was found that subsoil condition at each tower foundation were obviously different with regards to the average shear wave velocity of VS30 value and liquefaction potential.
2. According to the three dimensional effective stress analysis considering the specification of each foundation, it was suggested that significant amount of phase difference between foundations might occur not in only the foundation soil system shallower than about GL-20 m.
3. It was found that foundation of transmission tower behaved in sway motion with very small amount of rotation when pile tip was placed on relatively hard ground with shear wave velocity of more than 300 m/s.

For further study, the following may be issues for future to evaluate the earthquake response of transmission towers;

– To take wider area of geotechnical information with VS30 values, surface relief such as river banks into consideration for ground model.
– To consider upper bound and lower bound for soil properties, thickness of liquefied layer in foundation soil coupled analysis.
– To consider soil properties such as shear wave velocity profile of deeper parts of the ground more than GL-30 m, and to consider basin terrain and deep subsoil structure as well.

ACKNOWLEDGEMENT

The authors express their gratitude to TEPCO Power Grid, Incorporated for providing authors such valuable data and information related to subsoil and foundations of transmission towers.

REFERENCES

ADINA R&D, Inc. 2011: ADINA User Interface Command Reference Manual Volume 1: *ADINA Soils & Structures Model Definition.*

Disaster Management, Cabinet Office, Japan. 2001. *Report of Earthquake Damage Estimation due to the Tokai Earthquake, 10th conference, Nov.27th, 2001*, Handout 2–2: 11.

Iai S. 1993. Three dimensional formulation and objectivity of a strain space multiple mechanism model for sand, *Soils and Foundations*, 33(1): 192–199.

Iai S. 1995. Response of a Dense Sand Deposit during 1993 Kushiro-oki Earthquake, *Soils and Foundations*,.35(1): 115–131.

Iai, S., Tobita T., Ozutsumi O. and Ueda K. 2011. Dilatancy of granular materials in a strain space multiple mechanism model, *International Journal for Numerical and Analytical Methods in Geomechanics, Int. J. Numer. Anal. Meth. Geomech.* 35: 360–392.

Iai, S., Matsunaga, Y., Kameoka, T. 1992. Strain Space Plasticity Model for Cyclic Mobility, *Soils and Foundations*, 32(2): 1–15.

Iwasaki, T., Tokida, K., Tatsuoka, F. 1981: Soil Liquefaction Potential Evaluation with use of the Simplified Procedure, *Proc. 1st Intl. Conf. of Recent Advances in Geotechnical Earthquake Engineering and Soil Dynamics, St. Louis, Missouri*.

Ministry of Land, Infrastructure, Transport and Tourism. 2016, Search Engine for National Land Geoinformation "Kunijiban", http://www.kunijiban.pwri.go.jp/jp/index.html. (in Japanese).

Morita T., Iai S., Liu H., Ichii K., Sato Y., 1997. Simplified Method to Determine Parameter of FLIP, *Technical Note of the Port and Harbour Research Institute, Ministry of Transport, Japan*. (in Japanese).

National Research Institute for Earth Science and Disaster Prevention (NIED). "K-NET WWW service", Japan (http://www.k-net.bosai.go.jp/).

National Research Institute for Earth Science and Disaster Prevention (NIED). "Japan Seismic Hazard Information Station (J-SHIS)". 2016. (http://www.j-shis.bosai.go.jp/).

Ohta, H., Kawahara, A., Nakamura, H., Yamazaki, M., Hongo, E. 2014. Reproduction Analysis of Short-circuit between Transmission Lines during the 2011 Off the Pacific Coast of Tohoku Earthquake and a Consideration of the cause of the Event. *IEEJ Transactions on Power and Energy*, 134(8): 732–742. (in Japanese).

Yamanashi Prefecture, Japan. 2003: Survey of Deep Subsurface Structure Beneath Kofu basin. (in Japanese).

Yamanashi Prefecture, Japan. 2005: Report of Earthquake Damage Estimation due to the Tokai Earthquake in Yamanashi Prefecture. (in Japanese).

Yamanashi Prefecture, Japan. 2013: Report of Liquefaction Evaluation in Yamanashi Prefecture. (in Japanese).

Seismic Performance of Soil-Foundation-Structure Systems – Chouw, Orense & Larkin (Eds)
© 2017 Taylor & Francis Group, London, ISBN 978-1-138-06251-1

Seismic performance of a non-structural component with two supports in bidirectional earthquakes considering soil-structure interaction

E. Lim & N. Chouw
Department of Civil and Environmental Engineering, The University of Auckland, New Zealand

L. Jiang
National Engineering Laboratory for High Speed Railway Construction, Central South University, Changsha, China

ABSTRACT: Multiply-supported Non-Structural Components (NSCs) attached across different floors of a primary structure, e.g. building façades, advertisement boards, fire escapes, are particularly vulnerable to excitations in the vertical direction. To date, only very limited research has been conducted to address this issue, especially those using experiments representing more realistic scenarios. In this work, a physical model consisting of a four-storey primary structure and an NSC with two supports was considered to simulate a realistic interaction between the primary structure and NSC (PSSI). In the latter part of the experiment, the entire model was placed in a large box containing wet sand to simulate a more realistic support condition, i.e. Soil-Structure Interaction (SSI) was considered. This paper aims to investigate the simultaneous effect of PSSI and SSI on the seismic response of the NSC. The ground motion used was a real record measured at Kobe Port Island station in the 1995 Kobe earthquake. The ground motion was chosen due to the strong vertical motion that was anticipated to have significant effect on the NSC. The influence of bi-directional excitation and each of them individually on the response of the NSC will be showcased.

1 INTRODUCTION

Non-structural components (NSCs) include all non-load bearing component of an infrastructure that are typically attached to the primary load-bearing members (Adams, 2001; Villaverde, 1997). NSCs include a wide range of objects, e.g. building façade, advertisement boards, data acquisition system, and even household furniture. NSCs are particularly vulnerable during earthquakes because they are usually not designed to withstand dynamic loads (Lim and Chouw, 2014). In a seismic event, detachment and damage of NSCs as well as NSCs incurring impact loads on the primary structure and imposing risk to building occupants are realistic threats (Villaverde, 1997; Chen and Soong, 1988).

Proper seismic analysis of NSCs requires a thorough understanding of the interaction between the NSC and the primary structure, more commonly known as the primary-secondary structure interaction (PSSI) (see e.g. Naito and Chouw, 2003; Lim and Chouw, 2014). Developing a holistic analysis method for seismic design of NSCs is difficult because different types of NSC have different interaction with the primary structure, thus producing different response. Past numerical research on the dynamic response of NSCs including PSSI used many simplifications, e.g. only elastic linear structures and uni-directional horizontal excitations were usually considered (Igusa and Kiureghian, 1985a-c, Asfura and Kiureghian, 1986).

This research aims to establish an understanding of the seismic response of a multiply-supported NSCs attached across different floors of the structure. This type of NSC is expected to be more vulnerable to vertical motion rather than horizontal motion. In the experiment, the ground motion was applied in horizontal and vertical directions, individually and combined, to showcase the influence of each component on the response of the NSC. In the latter part of the experiments, the structure will be placed in a sand box to simulate a more realistic interaction between the primary structure and surrounding soil, i.e. soil-structure interaction (SSI).

SSI has been widely known to affect the response of structures in earthquakes. Although many past studies have suggested the benefit of considering SSI in the seismic analysis of primary structures (Mylonakis *et al.*, 2006; Qin *et al.*, 2013), not much has been explored on its effect on the response of NSCs (Lim et al., 2015). For NSCs with multiple supports, almost no research has been reported. Thus the main premise is to present the influence of bi-directional excitation on the response of the NSC including the simultaneous effect of PSSI and SSI.

2 EXPERIMENTS

The experimental model consists of two parts: (1) the primary structure, and (2) the non-structural component (NSC). To reveal the effect of SSI, the exact same model was fixed on the ground and placed in a sand box alternately. A 1.1 m deep wet sand in a 3 m × 3 m rigid box was used in the experiments. The same ground motions were applied in both cases. Figure 1 shows a sketch of the experimental model placed in a sand box including the scaled dimensions.

The structure is based on a four-storey prototype with a length scale of 1:4 and mass scale of 1:90. The modelling approach was adopted from Qin (2015) to produce correct horizontal and vertical forces without scaling the time. Thus, the prototype and model frequencies are the same.

The NSC is a slender frame with two supports attached across the top two floors of the primary structure, as shown in Figure 1. Compared to the beam the columns of the frame can be considered massless. The beam is made of steel with a mass (m_{NSC}) of 24 kg and size of 87 mm × 21 mm × 1740 mm. The natural frequencies of the component (f_{NSC}) are listed in Table 2. The mass ratio μ, and frequency ratio η, of the primary structure and NSC are also shown in Table 2.

The ground motion ($\ddot{u}_g$) considered in this research is a record of the 1995 Kobe earthquake measured at Kobe Port Island (KPI) station chosen due to its strong vertical motion. The excitation was applied in the horizontal (x), vertical (z), and horizontal and vertical (xz) directions. The horizontal excitation was applied in the direction of the weak axis of the primary structure.

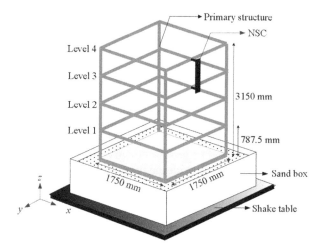

Figure 1. Experimental model placed in a sand box.

50

Table 1. Properties of the primary structure.

Properties	Prototype		Model
Inter-storey height	3.15 m		787.5 mm
Total height	12.6 m		3150 mm
Bay width	7 m		1750 mm
Floor mass (m_{PS})	29 tons		322 kg
Roof mass	24 tons		267 kg
Foundation size	$7 \times 7 \times 0.4 \text{ m}^3$		$1750 \times 1750 \times 10 \text{ mm}^3$
Fundamental frequency (f_{PS})	1.86 Hz (x axis)	6 Hz (y axis)	17 Hz (z axis)
Damping ratio	5% (expected)		

Table 2. Mass and frequency ratios between the primary structure and NSC.

$\mu = m_{NSC}/m_{PS}$	$\eta = f_{NSC}/f_{PS}$			
	Axis	f_{NSC} (Hz)	f_{PS} (Hz)	η
	x	200	1.86	$\eta_x = 111.10$
$\mu = 8.8\%$	y	17	6	$\eta_y = 2.83$
	z	8.6	17	$\eta_z = 0.51$

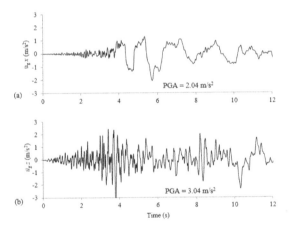

Figure 2. Scaled ground motions in (a) horizontal x and (b) vertical directions.

3 INFLUENCE OF BI-DIRECTIONAL EXCITATION ON A FIXED BASE STRUCTURE

The accelerations at the top and bottom of the NSC in x, y, and z directions were measured. The response presented in this section is based on the results from the tests on fixed base primary structure. Figure 3 shows the comparison of the acceleration at the top of the NSC in the x direction when one (x) and two-directional (xz) excitations were considered. As shown in Figure 3(a), considering both horizontal and vertical excitations gave an almost identical vibration shape compared to considering only the horizontal excitation. On the contrary, vertical excitation alone resulted in negligible response in the horizontal direction (Fig. 3(b)). As anticipated the NSC horizontal acceleration mostly due to the excitation in the horizontal direction.

On the other hand, the response in the vertical direction is mostly caused by the vertical ground motion. As shown in Figure 4(a), there was a significant increase in the vertical response when vertical excitation is introduced. However, when comparing the response due to vertical excitation alone and the bi-directional excitation, it is clear that considerable vertical response

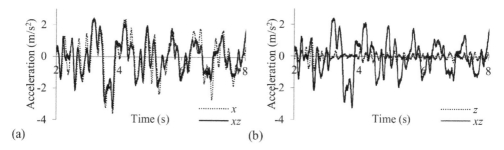

(a)
(b)

Figure 3. Effect of simultaneous x and z excitation on the NSC acceleration in the x direction.

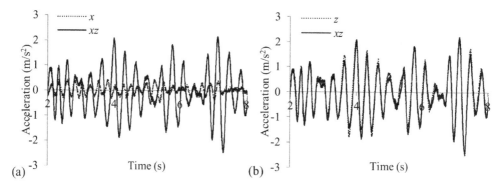

(a)
(b)

Figure 4. Vertical response in the NSC induced mainly by the vertical excitation.

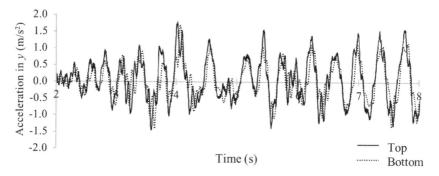

Figure 5. Negligible spatial coupling behavior within the NSC due to xz excitation.

occurred because of the vertical excitation. Current design provisions for NSCs (e.g. NZS 4219; ASCE 7-05) only consider horizontal excitations. The results of this research indicated that this might not be sufficient for all types of NSCs because the conventional consideration of only the horizontal excitation will not be able to produce a realistic vertical response. In the case considered, analyzing the response of the NSC due to individual x and z excitations will likely be sufficient.

Previous studies on the influence of multi-directional excitations performed by the same authors (Malligarjunan et al., 2015) reported that bi-directional horizontal excitations (xy) as well as three-directional excitations (xyz) increase the spatial coupling effect within the NSC. This was mostly reflected in the difference in the direction of the acceleration at the top and bottom of the NSC. Nevertheless, in the case considered, the spatial coupling behaviour of the NSC when xz was considered compared to one directional excitation is negligible, as shown from the accelerations of the NSC in y direction in Figure 5.

4 SIMULTANEOUS EFFECT OF PSSI AND SSI

Figures 6 and 7 show the acceleration at the top of the NSC in the x and z directions with and without considering SSI, respectively. The responses due to one directional excitation (x) are shown in Figures 6(a) and 7(a), while those due to two directional excitation (xz) in Figures 6(b) and 7(b). During the experiments considering SSI, the edge of the base of the primary structure uplifted against the sand surface. An impact occurred upon re-contact with the ground, inducing a sharp increase in the response of the NSC in the horizontal direction (the local maxima circled in Fig. 6). The peak accelerations during both tests occurred because of these sharp local maxima and were significantly larger than the corresponding fixed base tests.

The response of the NSC in the vertical direction was significantly larger when SSI was considered. As shown in Figure 7(a), when PSSI and SSI were considered simultaneously,

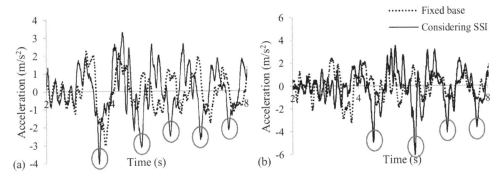

Figure 6.   Acceleration at the top of the NSC in the x direction due to (a) x, and (b) xz excitations.

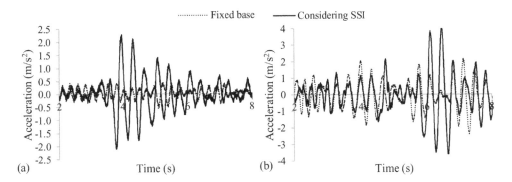

Figure 7.   Acceleration at the top of the NSC in the vertical direction due to (a) x, and (b) xz excitations.

Table 3.   Peak acceleration at the top and bottom of the NSC with and without considering SSI (m/s²).

Measured location		Peak acceleration due to excitation in					
		x		z		xz	
		Fixed base	SSI	Fixed base	SSI	Fixed base	SSI
Top	x	3.61	4.03	1.16	1.32	3.22	6.16
	z	0.54	2.27	2.49	2.67	2.46	4.04
Mid	x	3.21	3.38	1.80	2.30	3.67	4.67
	z	0.47	2.39	2.58	2.77	5.15	4.24

a strong vertical response was recorded even when the excitation was applied only in the x direction. When both directions of excitation were considered, the vertical acceleration was even more significantly increased. Most importantly, the maximum accelerations occurred at different time instants when different directions of excitation were considered.

The maximum accelerations at the top and bottom of the NSC with and without considering SSI are listed in Table 3. In all cases, the maximum accelerations are significantly higher when SSI was considered compared to the fixed base cases.

5 CONCLUSIONS

This paper reports the seismic performance of non-structural components (NSCs) with two supports considering soil-structure interaction through experiments. The structure was subjected to horizontal (x) and vertical (z) ground excitations, applied individually and simultaneously. Two support conditions were considered: 1) assumed fixed base and 2) considering soil-structure interaction. In the fixed base case, the horizontal excitation contributed to significant response in the horizontal direction, whereas vertical excitation to vertical response. Bi-directional excitation did not increase the response significantly. However, this finding proved that for certain types of NSCs, considering both directions of ground motions, at least individually, is required for more accurate predictions of the seismic response.

In the experiments considering soil-structure interaction (SSI), uplift of the primary structure occurred. The impact followed at the base of the structure due to the uplift induced sharp increase in the local maxima of the response in the horizontal direction. When PSSI and SSI were considered simultaneously, a strong vertical response was recorded even when only horizontal excitation was considered. When both directions of excitation were considered, the vertical acceleration was even more significantly increased. This indicates that neglecting SSI in designs of NSCs could potentially underestimate the seismic demand.

ACKNOWLEDGEMENTS

The authors thank the National Engineering Laboratory for High Speed Railway Construction of Central South University (China) for providing funding for the experiments under the Project no. 2013G002-A-1, and the Ministry of Business, Innovation and Employment through the Natural Hazards Research Platform (Award UoA3703249) for the scholarship to the first author.

REFERENCES

Adams, C. 2001. Dynamics of elastic-plastic shear frames with secondary structures: Shake table numerical studies. *Earthquake Engineering and Structural Dynamics* 30(2): 257–277.

ASCE 7-05, Chapter 13: Seismic design requirements for nonstructural components, minimum design load for buildings and other structures. *American Society of Civil Engineering Standard 2006*.

Asfura, A. & Der Kiureghian, A. 1986. Floor response spectrum method for seismic analysis of multiply supported secondary systems. *Earthquake Engineering and Structural Dynamics* 14(2): 245–265.

Chen, Y. & Soong, T.T. 1988. Seismic response of secondary systems. *Eng. structures* 10(4): 218–228.

Igusa, T. & Kiureghian, A.D. 1985a. Dynamic characterization of two-degree-of-freedom equipment-structure systems. *Journal of Engineering Mechanics, ASCE* 111(1): 1–19.

Igusa, T. & Kiureghian, A.D. 1985b. Dynamic response of multiply supported secondary systems. *Journal of Engineering Mechanics, ASCE* 111(1): 20–41.

Igusa, T. & Kiureghian, A.D. 1985c. Generation of floor response spectra including oscillator-structure interaction. *Earthquake Engineering and Structural Dynamics* 13: 661–676.

Lim, E. & Chouw, N. 2014. Effects of ground motion characteristics on structural response with primary-secondary structure interaction; *13th international symposium on structural engineering (ISSE-13)*.

Lim, E., Qin, X. & Chouw, N. 2015. Dynamic interaction of primary-secondary system considering soil-structure interaction; *6th Int. Conf. on Earthquake Geotechnical Eng.,* 1–4 November 2015. Christchurch: New Zealand.

Malligarjunan, R., Lim, E. & Chouw, N. 2015. Effects of multi-directional excitations on the response of a secondary structure with two attachments; *Proceedings of the Ninth Pacific Conference on Earthquake Engineering,* 6–8 November 2015. Sydney: Australia.

Mylonakis, G., Nikolaou, S. & Gazetas, G. 2006. Footings under seismic loading: Analysis and design issues with emphasis on bridge foundations. *Soil Dynamics & Earthq. Engineering* 26(9): 824–853.

Naito, K. & Chouw, N. 2003. Measures for preventing secondary structures from uplift during near-source earthquakes; *40 years of European Earthquake Eng.,* 26–29 August, Ohrid.

Qin, X. 2016. Experimental studies of structure-foundation-soil interaction effect on upliftable structure, PhD. Thesis, University of Auckland, Auckland, New Zealand.

Qin, X., Chen, Y. & Chouw, N. 2013. Effect of uplift and soil nonlinearity on plastic hinge development and induced vibrations in structures. *Advances in Structural Engineering* 16(1).

Standards New Zealand. Seismic performance of engineering systems in buildings—New Zealand (NZS 4219: 2009). *Wellington, New Zealand* 2009.

Villaverde, R. 1997. Seismic design of secondary structures: State of the art. *Journal of Structural Engineering* 123(8): 1011–1019.

Seismic Performance of Soil-Foundation-Structure Systems – Chouw, Orense & Larkin (Eds)
© 2017 Taylor & Francis Group, London, ISBN 978-1-138-06251-1

Seismic ground-structure interaction: A geotechnical practitioner's perspective

J.C.W. Toh
PSM, Sydney, Australia

ABSTRACT: This paper provides a geotechnical practitioner's perspective on seismic ground-structure interaction, in two parts. Firstly, a general discussion on ground-structure interaction in design practice is presented. Secondly, a bridge foundation design example is used to discuss design approaches, and highlight challenges faced in practice and how they might be addressed. It is concluded that for projects featuring complex seismic ground-structure interaction, structural and geotechnical work should be integrated as far as practical at all stages.

1   INTRODUCTION

This paper provides my perspective on seismic ground-structure interaction in engineering design practice. It is based on my ten or so years of experience in New Zealand and Australia, primarily as a geotechnical practitioner and secondarily as a researcher.

The first part of the paper is a general discussion on ground-structure interaction in practice, considering the spectrum of geotechnical and structural engineering design. It attempts to elaborate why seismic ground-structure interaction is a challenging field, and to explain some of the issues faced by practicing geotechnical engineers.

The second part of the paper presents an example of a foundation design that has recently been completed for a bridge in a high-seismicity area. Although it is a single example, it involves the same seismic ground-structure interaction issues that I have encountered in other projects, so it is a good basis for discussing design approaches. Note I've used the term ground-structure interaction as opposed to the workshop title term soil-foundation-structure interaction, only because the example design involves rock, not soil.

I hope that this paper provides some interesting insights on the issues that can be faced by geotechnical practitioners in this challenging field, which might further the understanding and collaboration with both structural practitioners and geotechnical academics.

2   DISCUSSION ON GROUND-STRUCTURE INTERACTION IN PRACTICE

2.1   *The different worlds of geotechnical and structural design engineers*

The worlds of geotechnical and structural engineering practice can often be quite different and separate. At a glance, structural designers can seem to operate within the relative comfort of established design codes, and enjoy the certainty of the properties of man-made materials; whereas much of a geotechnical designer's work revolves around unknowns, uncertainty, and risk, with little or no codified design methodologies. This means that while geotechnical and structural engineers can be quite comfortable working in their own respective worlds, unfortunately sometimes there is not always as much understanding or appreciation of the other world.

Where a design features geotechnical behaviour that is separate to structural response, this does not lead to any issues. However on some occasions, and especially in earthquake engineering, the two worlds of structural and geotechnical engineers collide, in a place called ground-structure interaction.

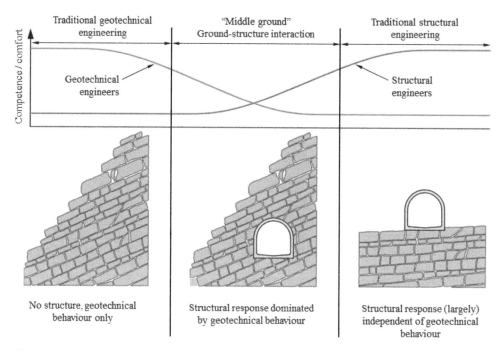

Figure 1. Spectrum of geotechnical and structural engineering design (Clarke, 2016).

Figure 1 was devised by my colleague (Clarke, 2016), and I find it an excellent way of explaining these ideas. Although the figure relates to statics and tunnels, the same ideas apply for earthquakes and foundations. Regarding Figure 1 and ground-structure interaction the question is; who looks after the "middle ground"?

Sometimes it can be easier for geotechnical engineers to cover the "middle ground", perhaps because it is the geotechnical behaviour that is an input to the structural response rather then vice versa. Of course this doesn't have to always be the case. Often both the geotechnical and structural engineers need to be fully aware of the interaction issues faced in the "middle ground" to arrive at the optimal design solution, especially in more complex designs. In any case, working in the "middle ground" requires some knowledge of the other world.

2.2 *Geotechnical challenges*

Some challenges for a geotechnical engineer working on an aseismic foundation design and venturing into the "middle ground" can include:

– Understanding the structure. In the absence of ground-structure interaction, what is its behaviour? What aspects are 'governing' the structure design? Is it more desirable to have a rigid system to minimise deflections, or a flexible system to minimise structural actions?
– Understanding how the ground-structure interaction affects the dynamic response of the structure. What geotechnical aspects have a significant effect on the structure response? What are conservative assumptions for the ground response? Are these assumptions unconservative for the structure response? And vice versa.
– Defining the interfaces between geotechnical and structural disciplines.
– Deciding what analysis tools to employ. What methodologies and software are readily available and accessible? Are pseudo-static methods sufficient or is dynamic analysis required? Can the behaviour be adequately bound?
– How to achieve and demonstrate compliance with relevant design codes and manuals, and other requirements, given often there is limited precedent and a lack of established or prescribed methodologies?

– How to develop an economic design avoiding excessive conservatism? How much conservatism should be adopted at each step along the way? What is the cumulative effect of this? Does the end design reflect 'real' behaviour?

The last four challenges listed can be unique to practice, as they are not purely technical considerations and therefore not necessarily an issue in research as well. I'm sure there are equivalent challenges for a structural design engineer.

3 EXAMPLE OF BRIDGE FOUNDATION DESIGN IN HIGH SEISMICITY AREA

3.1 *Purpose of example*

This second part of the paper presents an example foundation design for a bridge in a high seismicity area in New Zealand. The design was undertaken by separate geotechnical and structural engineering consultants, working in conjunction. Although it is a single example, it involves many of the seismic ground-structure interaction issues that I have encountered in practice, so it is a good basis for discussing design approaches, and highlighting challenges and how they might be addressed. It is emphasised that the approaches presented are one way, and not the only way, to undertake such a design. Because the purpose of the example is to discuss design approaches, and due to space constraints, full technical details of the seismic hazard, geotechnical model, structure design, geotechnical analysis, etc, are unfortunately not provided.

3.2 *Background*

The design was undertaken following the requirements of the New Zealand Transport Agency Bridge Manual (2014). This document is based on limit state design but in a geotechnical sense it is sufficiently flexible to permit the type of innovative design approaches that are often required for unique structures in high seismicity areas. In particular, geotechnical analysis and design methods are not overly prescriptive, and permanent deformations are acceptable under many of the limit states provided certain performance levels are maintained.

3.3 *Seismicity*

Bedrock horizontal peak ground accelerations for limit states relevant to the geotechnical design are:

– 0.20 g for the serviceability limit state (SLS).
– 0.68 g for the ultimate limit state (ULS)
– 0.83 g for the maximum credible earthquake (MCE).

The structure performance requirements for each of these events are essentially: generally undamaged structure at SLS, damage at ULS limited to easily replaceable elements, with such damage not to preclude immediate access for emergency vehicles, and no collapse at MCE.

3.4 *Ground conditions*

The bridge spans a stream that has caused a valley incision into surrounding flat topography. The slopes forming the sides of the valley are at angles of between 30° and 45°; with no identified active landslides of significant scale identified. Bedrock is near or at the surface. The rock mass is a fractured interbedded sandstone and mudstone with significant tectonic disturbance so sheared zones are frequent. Primary defect sets are steeply dipping. Figure 2 indicates the rock mass at founding level; the rock mass above this is more weathered and fractured.

Beyond the usual difficulties in assessing static rock mass strength and modulus (which were investigated by a combination of laboratory testing and in situ geophysical testing), key design considerations were dynamic rock mass strength and whether the rock mass would behave in a ductile or a brittle manner post-yield, and dynamic rock mass modulus.

Figure 2. Rock mass at founding level (notepad indicates scale).

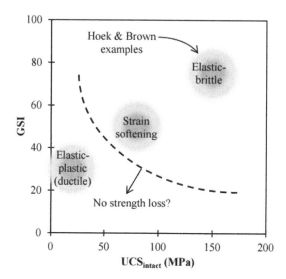

Figure 3. Consideration of post-yield behavior of rock mass.

There is little in the literature for post yield rock mass strengths under static conditions and even less for dynamic conditions. A GSI (Geological Strength Index) based approach (Hoek & Brown, 1997) predicts that high strength good quality rock masses are brittle and experience significant post yield strength loss, while low strength fractured rock masses are ductile and do not experience strength loss post yield. Refer to Figure 3 for an interpretation of this assessment.

With the rock mass at founding level having a GSI of around 30 to 50, and an average UCS of around 60 MPa, and these parameters decreasing at lesser depths, the rock mass was assessed to be ductile (i.e. no strength loss post-yield) for the purposes of calculating coseismic displacements of large and medium scale rock mass failure mechanisms using Newmark sliding block methods (e.g. Newmark, 1965). This assumption of ductility did not extend beyond a certain amount of displacement, or to other mechanisms e.g. small scale mechanisms and those mechanisms controlled by geological structure.

3.5 *The structure*

The structure and foundation arrangement was selected based on the site geometry, constructability, and ability to satisfy the design criteria including the seismic performance requirements described above. The bridge is about 230 m long and has three spans across two piers (Figure 4). The superstructure comprises four steel open-top box girders which act compositely with a precast prestressed concrete deck.

The pier columns are reinforced concrete and are hollow rectangular in section. They are fixed into foundations near the toes of the valley slopes (i.e. a full moment-resisting connection). The foundations comprise reinforced concrete footings with passive ground anchors to provide additional resistance to moment and lateral loads (Figure 5). The bridge deck sits on pot bearings atop the pier columns which fully transfer vertical and lateral loads but not moments. The optimal combination of span and pier column height was selected based on performance of the structure.

The abutments comprise reinforced concrete beams fixed into reinforced concrete pile foundations located at the crests of the valley slopes. At the abutments, the bridge deck sits on base isolator lead-rubber bearings atop the beams, which transfer vertical loads, provide lateral stiffness to resist lateral loads up to a certain point (beyond which, e.g. during large seismic loading, sliding will occur), but do not transfer moments and so decrease structural actions.

Bearings at both the abutments and piers were designed with the ability to be reset following an earthquake.

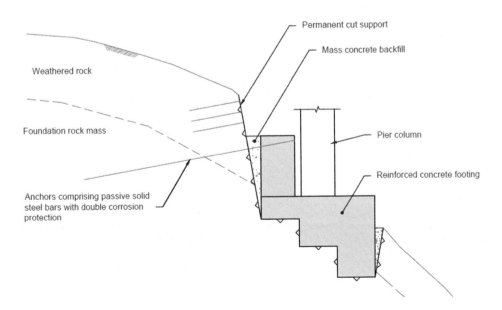

Figure 4.   A perspective view of the structure.

Figure 5.   Elements of pier foundation design.

61

The structural modelling incorporated foundation elements, and represented ground-foundation stiffness using yielding springs assessed using the methods described below.

3.6 Development of design considering ground-structure interaction

The design was developed iteratively with regards to ground-structure interaction because:

− The foundation design determines foundation stiffness, which affects superstructure actions and deflections, which in turn affects foundation design. There is a need to control both structural actions and deflections but these are achieved by the opposite goals of decreasing and increasing foundation stiffness respectively. Ultimately an optimum balance needs to be achieved.
− Provision of sufficient foundation capacity, both globally and of individual elements, depends on sources of loads and the coincidence of the loads. Further, in achieving sufficient capacity, the foundation stiffness and therefore the superstructure actions, can be affected. The details of these loads change as the design develops and the interaction is resolved.
− Slope loads imposed on foundation elements depend on how much the foundations displace as a result of loading from the superstructure.
− Different seismic limit states have different performance requirements, and at the outset it is not clear which limit state will 'govern' the design.

Due to such an iterative process it was important to identify key design aspects and constraints at an early stage. Once the interaction was understood at a high level, the details of the design could be progressed.

3.7 Foundation stiffness

The NZTA Bridge Manual includes the following requirements in relation to foundation stiffness: "*When estimating foundation stiffness to determine the natural period(s) ... a range of soil stiffness parameters ... shall be considered*", and "*Upper and lower bound strength and stiffness properties of the soils shall be applied in order to assess the most adverse performance likely of the structure, which is to be adopted as the basis for its capacity design*".
Figure 6 shows the design ULS response spectrum. Modal analysis by the structural engineer assessed a "fixed base" superstructure period in the order of 2 seconds for the ULS and MCE. Similar analysis incorporating "lower bound" foundation stiffness (see later for geotechnical method of estimating foundation stiffness) assessed a superstructure period in the order of 2.5 seconds. Because the "fixed base" period lay on a relatively "flat" part of the response spectrum, adopting "lower bound" foundation stiffness only decreased superstructure actions by about 10%. Observing this relative insensitivity, it was decided to capture the range of possible behaviour by using "fixed base" models to assess structural actions, and models with "lower bound" foundation stiffness to assess structural displacements. This simplified the interfaces

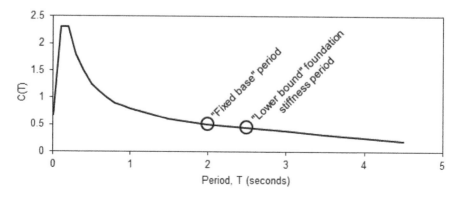

Figure 6. Elastic design site hazard spectrum for horizontal loading, and bridge fundamental period.

and iterations between the geotechnical and structural designers and meant that the foundation could be sized to achieve adequate capacity, and design iterations could be undertaken to assess how stiff the foundation was required to be to control displacements of the superstructure and bearings, and assess the optimum balance between these aspects.

3.8 *Foundation capacity*

The NZTA Bridge Manual includes the following requirements in relation to foundation capacity: "*Foundations shall be designed for bearing capacity and stability to resist combined horizontal and vertical loadings with acceptable displacements and settlement*", and "*Foundation design shall be based on appropriate sound design methods*". A load and resistance factored (LRFD) design approach is prescribed for overall foundation capacity. However localised yielding is not precluded, which is good because sometimes the quest to achieve no yield can lead to extremely stiff foundations that attract more load. A side effect of this is that localised yielding can result in permanent foundation displacements even if capacity is adequate. The Bridge Manual also emphasises the importance of ductile behaviour once capacity is reached. Implementing LRFD design usually presents difficulties for geotechnical engineers, let alone when it is implemented for seismic ground-structure interaction designs.

3.9 *Abutment foundation analysis*

Limit equilibrium analyses were used to assess the ultimate lateral capacity (passive resistance) of the abutment piles; the slope adjacent to the piles and the varying rock mass meant that analytical solutions were not suitable. The piles needed to have sufficient lateral capacity to accommodate the shear loads imposed by the superstructure.

Following Bridge Manual requirements slope stability analyses were undertaken using limit equilibrium techniques, and where factor of safety was less than one coseismic displacements were assessed using methods based on the Newmark sliding block method. These displacements were then used to estimate slope loads imposed on the piles, considering interaction. Noting that the displacements represent the permanent coseismic displacements accumulated over the duration of an earthquake (mainly during peaks of downslope acceleration), this method is fundamentally incompatible with assessment of dynamic behaviour.

The results of dynamic analysis of the structure were therefore used to assess the coincidence of slope movements (and resulting slope loads imposed on piles) with shear loads applied to the ground by the superstructure via the piles (Figure 7). Because the superstructure had a reasonably long period, the loads applied to the ground by the superstructure did not coincide with the peak ground accelerations at which the slope may impose loads on the piles. This avoided excessive conservatism that could have resulted from inclusion of additional loads in the analyses, and is an example of the geotechnical design benefitting from use of structural analysis.

3.10 *Slope mechanisms interacting with pier foundations*

Figure 8 shows some slope mechanisms that interact with the pier foundations. Large scale 'global' mechanisms had factor of safety exceeding one for the design earthquakes.

Some medium scale mechanisms intersecting the side of the foundations had a factor of safety less than one for the larger design earthquakes. Limit equilibrium analyses were used to assess slope displacements and potential loads imposed on the foundations by these mechanisms. In the geotechnical foundation analysis (see below), these loads were only applied coincidentally with superstructure loads if the slope displacement exceeded the displacement of the foundation due to the superstructure loading, thereby avoiding excessive conservatism.

Surficial mechanisms involving small volumes of soil or weathered rock were addressed firstly by using the top of the foundation to physically engage as much of the slope as possible thereby minimising height of permanent cut, and secondly stabilising the remaining permanent cut with permanent rock bolts and shotcrete. Analysis was used to show that the pier columns had adequate capacity to accommodate rock fall impact loads.

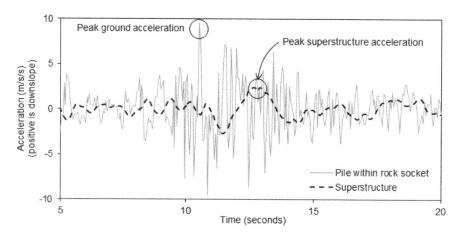

Figure 7.   Non-coincidence of ground accelerations with abutment pile loads applied to ground.

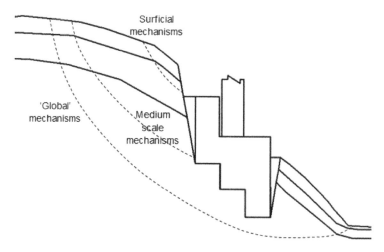

Figure 8.   Slope mechanisms affecting pier foundations.

3.11   *Pier foundation geotechnical analysis*

Geotechnical numerical modelling of the pier foundations was undertaken to assess foundation stiffness and demonstrate foundation capacity (context of both discussed above). The stiffness estimates were used as inputs to the structural models.

The numerical model is shown in Figure 9 and was a two-dimensional static finite element model, with the elastic-plastic ground model including rock defects, and rock mass or rock substance between the defects. The two-dimensional simplification was considered slightly but not overly conservative given the fractured rock mass. However it was still important to consider sensitivity of foundation performance to rock structure, particularly defect orientation. The numerical model did not directly include slope mechanisms which were assessed separately by the prescribed limit equilibrium methods as described above.

For each design scenario, the following analysis stages were adopted to provide the outputs required to demonstrate capacity and provide stiffness:

1. Establish equilibrium state for the existing ground and slope profile.
2. Excavate ground required to construct the foundation.
3. Install foundation including footing and anchors. Adopt design (factored) strengths for the anchors as opposed to ultimate strengths.
4. Incrementally apply the vertical static load from the superstructure.

64

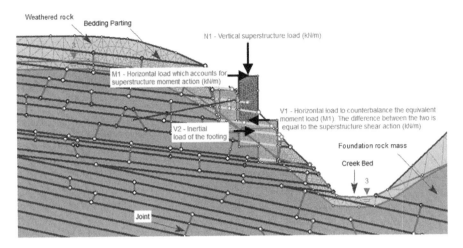

Figure 9.   Numerical model for examining pier foundation behaviour.

5. Incrementally apply the shear and moment seismic design (factored) loads imposed on the foundation by the superstructure, as well as the inertial load of the footing and any slope loads on the foundation. The results from this stage of analysis were used to assess load versus displacement characteristics and from this foundation stiffness was assessed, which was an input to the dynamic structural models.

6. Reduce the strength of the ground including rock mass, rock substance, and rock joints, by the prescribed geotechnical strength reduction factors. Geotechnical foundation capacity requirements were deemed to be satisfied if convergence of the model occurred at the final stage of analysis. Capacity was also verified by free body calculations. Mechanisms captured in the modelling and interrogated included bearing, overturning, sliding along footing-ground interfaces, sliding through the ground beneath the foundation, and ground anchor failure mechanisms including tensile failure of the anchor, pullout along interfaces, failure of the surrounding ground, and rupture of corrosion protection ducts.

7. A separate elastic model was analysed in parallel, with residual permanent foundation displacements following an earthquake calculated as the difference between the elastic and elastic-plastic model results. These permanent displacements occur because although 'global' capacity is satisfied, local yielding is permitted to occur for example in local areas of highly stressed ground. Although acceptable geotechnically, these permanent deformations also had to result in acceptable effects on the superstructure. It was noted that these permanent displacements are an upper bound because in reality earthquake motion involved alternating acceleration in opposite directions rather than a sustained unidirectional load.

It was considered that static analysis was required; dynamic analysis probably would not have been able to provide the information in the same form as described above. Further, there were anticipated significant issues with implementing a LRFD approach based on dynamic analysis, e.g. reducing strengths and increasing loads were expected to give misleading results not representative of real behavior. A dynamic analysis would however have been able to investigate behaviour beyond the point of non-convergence of the static analysis (i.e. where the system is temporarily out of dynamic equilibrium).

3.12   Interaction of geotechnical and structural foundation analysis and design

Figure 10 summarises the interactions between the various analyses and designs undertaken by the geotechnical and structural engineers. Implementing the multiple interactions and feedback loops was easier using a series of simple analysis tools (as opposed to a single complex model).

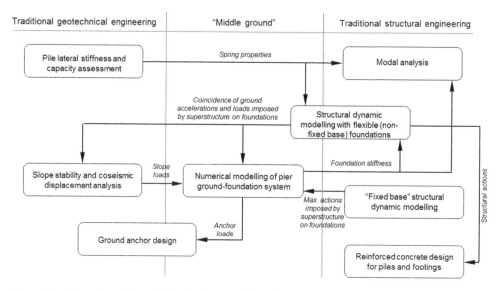

| Traditional geotechnical engineering | "Middle ground" | Traditional structural engineering |

Figure 10.   Interaction of geotechnical and structural foundation analysis and design in presented example.

4   CONCLUSIONS

In design projects where there are complex seismic ground-structure interaction aspects, it is imperative that both the structural and geotechnical engineers have a good understanding of each other's work including inputs, analytical methods, and outputs. Early integration of structural and geotechnical aspects of a design assists in understanding constraints and developing a satisfactory concept. In completing the details of the design, ongoing communication is also very beneficial, and there can be a multitude of ground-structure interaction considerations to resolve. It is possible and probably beneficial, for geotechnical engineers to make use of analysis by structural engineers, and vice versa, to complement and supplement their own analyses.

These conclusions are demonstrated, in relation to the presented example, by Figure 10. Considering the interactions between the structural and geotechnical engineering analyses and designs shown in Figure 10, it is clear that both structural and geotechnical engineers need to venture into the "middle ground" shown in Figure 1.

It is also worth noting, in comparison to research, that seismic ground-structure interaction in practice presents some unique issues to be addressed including defining interfaces between disciplines, selection of analysis tools, demonstration of compliance with standards, and development of an economical design. These issues are additional to purely technical considerations.

REFERENCES

Clarke, S.J. (2016) Personal communication.
Hoek, E. & Brown, E.T. (1997) Practical estimates of rock mass strength. Int. J. Rock Mech. Min. Sci. 34(8).
New Zealand Transport Agency (2014) Bridge manual SP/M/022 Third edition, Amendment 1.
Newmark, N.M. (1965) Effects of earthquakes on dams and embankments. Geotechnique 15(2).

Seismic Performance of Soil-Foundation-Structure Systems – Chouw, Orense & Larkin (Eds)
© 2017 Taylor & Francis Group, London, ISBN 978-1-138-06251-1

Dynamic lateral load field testing of pile foundations to determine nonlinear stiffness and damping

L.S. Hogan, M.J. Pender & L.M. Wotherspoon
Department of Civil Engineering, University of Auckland, Auckland, New Zealand

ABSTRACT: Currently, there exists a limited amount of in situ experimental data to quantify the non-linear lateral stiffness and damping of pile foundations, yet these properties are key inputs into all pile design procedures. Presented in this paper are the outcomes from a large scale experimental field testing program used to characterise these properties on three screw piles installed in stiff clay soils. The test piles extended 1.25 m above ground and the supported mass ranged from 1.4 to 3.2 tonnes. Testing was completed in three phases. First, monotonic quasi-static lateral load was applied above ground level in order to quantify the nonlinear force-displacement behavior of the soil-pile system. Second, at the desired maximum load of each static phase, the load was released suddenly and the pile was allowed to undergo free vibration to quantify the damping inherent in the system. The maximum static loads were incrementally increased for each test pile to assess the effect of gap development and compressive failure of the soil around the pile on the dynamic response. Finally, the piles were subjected to pseudostatic cyclic loading after the soil around the pile had been allowed to recover for some months, and the response was compared to the initial pullback and snapback testing. A summary of the test setup, loading regime, and results from the testing are presented. The most important conclusion from the paper is that the equivalent viscous damping ratio varies with the magnitude of the pullback force and is displacement dependent. Although the elastic damping of the piles was found to be about 3%, even for displacements as small as 1% of the pile shaft diameter the damping ratio was found to be 10% or more.

1 INTRODUCTION

The use of pile foundations in seismically active zones requires an understanding of the non-linear stiffness and damping provided by the foundation in order to assess the seismic demands on the supported structure. Currently there has been limited dynamic lateral testing of pile foundations, and in particular screw piles in which the pile is installed by using a helix at the pile tip. Testing that has been reported has tended to focus on small displacement dynamic response, (Blaney and O'Neill 1986, Boominathan and Ayothiraman 2006), and there has been little work to quantify the damping provided from pile foundations when the lateral deformations are sufficient to generate significant nonlinear pile-soil interaction (Cox et al. 1974, Reece and Welch 1975, Anderson et al. 2003, Janoyan et al. 2006). Yet the lateral loading of long piles almost always involves nonlinear soil-pile interaction at all but loads very small in relation to the lateral capacity of the piles. It is expected that the damping increases with lateral displacement but that this might be countered by the formation of gaps between the upper parts of the pile shaft and the surrounding soil.

Outlined in this paper are the results of a testing program that investigated the response of closed-end, hollow steel tube screw piles. A set of three piles were subjected to static monotonic, slow cyclic loading, and dynamic snapback testing to quantify free vibration response. The response of the piles at increasing horizontal loads was assessed, and the different loading regimes compared against one another.

2 TEST DESCRIPTION AND SETUP

2.1 *Site characteristics*

The site is located approximately 30 km north of Auckland in a region noted on the relevant geological map as part of the Northland Allochthon (Edbrooke 2001). The surface and near surface soils at the site are stiff clays, LL 60% and PL 31% with natural water content towards the plastic limit. The soil conditions at the site were determined prior to pile installation with CPT soundings at the centerline location of each pile and a borehole adjacent to the piles. Both CPT and borehole depths extended down to a depth of 8 m, which was the approximate pile depth. Borehole cores and CPT interpretations using soil behavior type index from Robertson and Cabal (2010) characterised the site as consisting of stiff silty clays overlaying very weak mudstone.

The tip resistance and friction ratio from the CPT soundings of each pile, SP1, SP2, and SP3 are shown in Figure 1. As the soil conditions at the site are fairly consistent the CPT results for each pile are similar, with an average tip resistance of 2 MPa for the top three metres and an average friction ratio of 6–7% over the same depth. Based upon interpretations from Robertson and Cabal (2010) the average Young's modulus of the soil for the top 3 m was approximately 40 MPa. The large tip resistance apparent at SP1 from the ground surface to approximately 0.5 m likely results from the presence of fill placed on site for the concrete driveway 3 m from the pile installation. The presence of this fill material into the test site was also identified from the borehole log.

In Figure 2 profiles of water content and shear wave velocity with depth are given. The water contents defined through laboratory testing were obtained from the cores obtained from two boreholes during the site investigation work. The shear wave velocity profile was obtained using surface wave techniques, with the profile representative of the overall site.

2.2 *Test setup*

The test piles, denoted as SP1, SP2, and SP3, were 220 mm diameter hollow steel pipe with capped ends and a 550 mm diameter helix at the base. The piles were installed in a line to a depth of 7.75 m to achieve long pile behavior in which pile deformation and rotation was restricted to an active length near the ground surface. The pile extended 1.25 m above ground with a free head condition. A 4.0 m spacing was maintained between the piles in order to minimize pile-to-pile interaction. Steel billets were installed on top of each pile to bring the dynamic response of the piles within the frequency range of interest of 2–4 Hz ($T_n = 0.25$ to

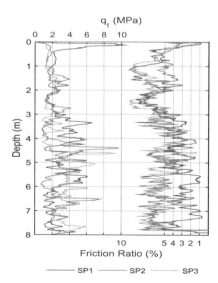

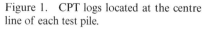

Figure 1. CPT logs located at the centre line of each test pile.

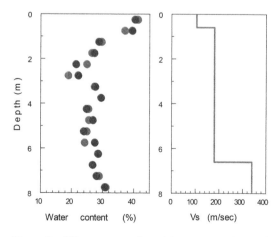

Figure 2. Water content from laboratory testing and shear wave velocity profiles.

68

0.5 s) as this range corresponds with the natural period of the structures that are typically supported by this size pile. Pile SP1 had approximately 1.4 tonnes of mass loaded on its head and piles SP2 and SP3 both had approximately 3.2 tonnes of mass each. The mass was connected to the pile head as shown in Figure 3.

Soil was excavated around the pile to a depth of approximately 100 mm and filled with water. The water was removed the day of testing to help maintain consistent water content between different pile test set ups.

Piles were loaded using a hydraulic jack connected between two piles. The two piles were pulled towards each other and deformations were measured from an independent reference frame with anchor points 2 m from the pile center line to minimize interaction during dynamic response (galvanized trusses in Figure 3). Horizontal displacement was measured at ground level, load height, and centre of mass. Rotation at the ground level was calculated from curvature measurements using portal gauge extensometers on either side of the pile, and accelerometers were installed at the centre of mass both in the direction of loading as well as in the orthogonal direction to capture accelerations during snapback testing. A schematic of the test setup and instrumentation locations is provided in Figure 4. The data logging system used a 16 bit A/D converter and during the dynamic snapback response the channels were logged at 1000 Hz.

2.2.1 Loading protocol

Lateral pile testing occurred in three phases. The first phase consisted of a force-controlled monotonic pullback in which two piles were pulled towards each other to characterise their nonlinear static lateral stiffnesses. Once the desired pullback force had been reached, the load was released using a quick-release shackle. During this snapback phase, the two piles were allowed to undergo free vibration to quantify the dynamic response of both piles. The center pile (SP2) was then pulled in the opposite direction to the same load level repeating both pullback and snapback phases. As such, the exterior piles, SP1 and SP3 were loaded in only one direction while the center pile, SP2, was pulled back and forward in two directions. After completing the pullback and snapback for a given load level, the loading protocol was repeated for an increased load level. Load steps for the monotonic pullbacks of phase one and two of testing are provided in Table 1. The load steps for monotonic pullback testing were only used once for each direction of the piles.

Before and after each load increment, the depth of gapping of the soil surrounding each pile was determined based on the elastic dynamic response of the pile and the surrounding soil, with the change in pile length calculated using the change in frequency from the equivalent single

Figure 3. Typical Test Pile (SP2) showing billet attached. (On either side reference frames are seen.).

69

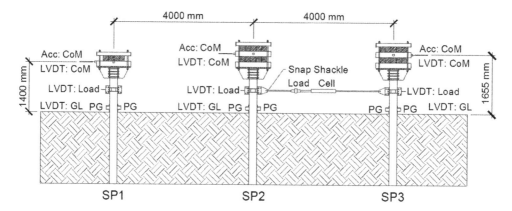

Figure 4. Schematic of test set up and typical instrumentation per pile. Acc: CoM = accelerometer at centre of mass, LVDT: CoM = LVDT at center of mass, LVDT: GL = LVDT at ground level, PG = portal gauge extensometer.

Table 1. Load steps for monotonic and slow cyclic testing of piles SP1, SP2, and SP3.

Monotonic pullbacks[1] (kN)	Slow cyclic (kN)	Cycles per load step
5	5	3
10	10	3
20	20	3
40	40	2
60	60	2
–	75	1

[1]only 1 cycle per load step was performed during pullback/snapback testing.

degree of freedom oscillator. The pile was struck with hammer blows to excite the pile and mass, a soft headed sledge hammer was used to avoid high frequency contamination of the response.

Three months after the initial testing, the piles were retested with a slow two-way cyclic protocol for comparison between the monotonic pullback and dynamic snapback response to this more traditional testing method. The slow cyclic testing was also performed to investigate the ability of the soil surrounding the pile to recover after three months of wet weather and whether the same lateral stiffness that was observed during the initial phase of testing was evident. The time interval of three months was chosen to coincide with available site access and previous experience with soil recovery from previous testing (Sa'don et al. 2014). The loading protocol for the slow cyclic testing was similar to the monotonic pullback testing in which two piles were pulled towards each other at a given load level, but during slow cyclic testing, the load was released in a slow and controlled manner. Both exterior piles were again only loaded in one direction, and the centre pile was loaded in two directions. Unlike the monotonic loading phase, a series of cycles were performed at all but the highest load increment to compare to the multiple low level cycles and potential soil damage present during the snapback phase of testing. The load steps used during the slow cyclic phase of testing and the number of cycles at each step are shown in Table 1.

3 STATIC PULLBACK RESULTS

3.1 *Static and slow cyclic pull-back tests*

The force-displacement response of the pullback phases of the testing of Pile SP1 is shown in Figure 5 and in Figure 6 there is the result of the slow cyclic loading of Pile SP2 loaded

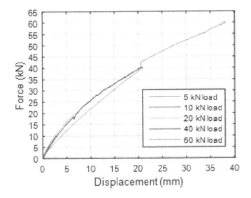

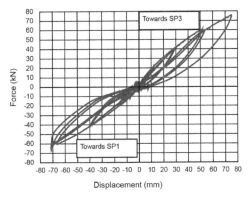

Figure 5. Force-displacement pullback response of SP1. Displacement measured at load height.

Figure 6. Force-displacement hysteresis of Pile SP2 during slow cyclic loading. (Displacement measured at load application height.).

alternatively in the direction of Pile SP1 and then SP3. Figure 5 shows the characteristic non-linear response of the pile to static lateral loading. The small jump in force during the 60 kN loading appears to be an instrumentation error. A small amount of degradation is apparent with increasing numbers of cycles. Figure 6 shows how the unloading portions of the curves follow a different nonlinear path and hence how there is work expended during a complete cycle of loading and unloading. Under dynamic response this is the mechanism that provides hysteretic damping.

One aspect of the pile-soil interaction that contributes to the nonlinearity is the formation of gaps between the pile shaft and the surrounding soil. After each load cycle a simple method (inserting a length of metal measuring tape) was used to estimate the gap depth. Gap depths between 0.25 and 2 pile diameters were indicated. Another approach to estimating gap depth is discussed below.

The slow cyclic load results shown in Figure 6 were obtained three months after the initial pullback testing; our intention was to see to what extent the soil around the pile "healed" during the rest period. The loading portions of hysteretic response for both piles displayed behavior similar to that observed during the initial pullback testing portion summarized above. The response to loading below 10 kN was essentially elastic and once loading exceeded 20 kN there was a noticeable reduction in stiffness for subsequent cycles at the same load level until the previous maximum displacement was exceeded.

A small amount of residual displacement of approximately 15% of the maximum deformation at each cycle was observed during the slow cyclic testing. Because this residual displacement was observed at small load levels it is unlikely a result of pile yielding. More likely, the residual displacement is due to soil filling the gap that formed behind the pile during loading and as such stopped the pile from returning to zero displacement. This characteristic was visually observed during testing.

4 DYNAMIC TEST RESULTS

After each dynamic snapback test, a hammer impact test was used to provide low-level excitation so that the elastic natural period of the pile system could be observed. Figure 7 shows the response of the accelerometer at the centre of mass for three hammer blows after a snapback from a 10 kN pullback force. Also shown in Figure 7 are envelopes used to characterise the damping of the pile response from the hammer blow decays. In this case equivalent viscous damping ratio of about 3% is indicated for the three hammer blows analysed using logarithmic decrement (Chopra 2007).

The frequency content within the accelerometer traces in Figure 7 was obtained using a Fast Fourier Transform. For each hammer blow sequence data from at least three recordings were processed. In all cases there was a clearly defined first mode frequency of the system which decreased with the magnitude of the pullback force that was applied prior to the snap-back, as shown in Figure 8. As explained above the decrease in natural frequency is assumed to be a consequence of increasing depth of the gapping between the pile shaft and the surrounding soil as the pullback force is increased.

Figure 9 shows the dynamic snapback response of pile SP2 when released from a 60 kN pullback in the direction of pile SP1; it is apparent that there is considerable damping in the system during the free vibration decay. In Figure 10 the data from Figure 9 has been overlaid with the calculated response to a sudden removal of load for a single degree of freedom model of the system, (Inman 2014). It is apparent that an equivalent viscous damping ratio of 20% matches the measured response very well for most of the first cycle of response, after which the damping in the system decreases. Using the logarithmic decrement method the relationship between damping per cycle and mean of the two displacement peaks is plotted in Figure 11 for snap data from pull back forces between 5 and 60 kN. In this figure the displacements are normalized with respect to the pile shaft diameter. The most important observation from is that the damping value is not constant but varies with the magnitude of the pullback force,

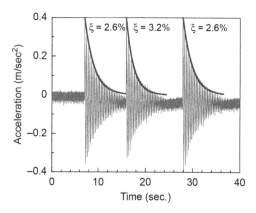

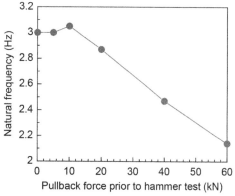

Figure 7. Pile acceleration response to hammer blows following the snapback from a 10 kN pullback.

Figure 8. Shift in the elastic natural frequency of the pile system after each load step characterised using hammer strike testing.

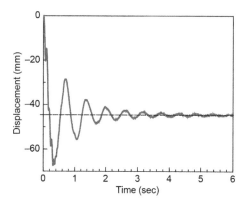

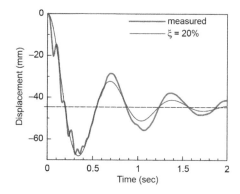

Figure 9. Displacement response of Pile SP2 following the snap release from a 60 kN pullback force.

Figure 10. Snapback response of Pile SP2 fitted to the elastic response from step function excitation to determine equivalent viscous damping.

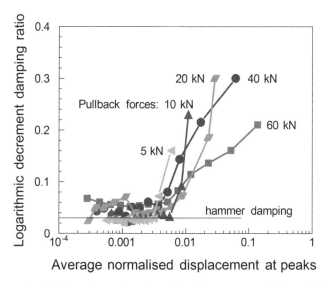

Figure 11. Observed equivalent viscous damping ratio with respect to pile lateral displacement.

or the mean of the displacements corresponding to the two peaks for which the logarithmic decrement is calculated. There is also a suggestion, looking at the result for the pullback from 60 kN, that as the pullback force increases a point is reached where there is so much degradation of the soil around the pile shaft that the maximum damping value decreases. Also plotted in Figure 11 is the damping value obtained from the hammer impulse tests, which is seen to be at the lower limit of the damping characterised during the snapback testing.

5 CONCLUSIONS

Three long 220 mm diameter hollow steel piles were tested in stiff residual clay and were subjected to monotonic, cyclic, and snapback loading. While the piles were displaced to as much as 20% of the pile diameter, nonlinearity was observed even at lateral displacements of a few percent of the pile shaft. As a consequence of the nonlinear response of the soil there is a continuously decreasing secant lateral stiffness for the pile shaft as the lateral load is increased. This decrease in secant stiffness does not appear to degraded with additional cycles and as such, monotonic backbone curves appear to be sufficient to capture the lateral response of the pile.

The post-snapback hammer impulse tests showed a reduction in natural frequency of the pile—mass system as the pullback force prior to the snapback increases (Figure 8), which is controlled by the increase in the depth of soil gapping that develops adjacent to the pile shaft. These small displacement impulses also revealed that the equivalent viscous damping of the pile was observed to be about 3% (Figure 7), which provides a measure of the elastic damping of the system similar to the global damping ratio of a structure usually assumed by equivalent static methods of building or bridge design. For further larger displacements, it was found that the observed equivalent viscous damping increased to 5% for displacements as small as 1% of the pile shaft and as large as 30% for pile shaft displacements of 8%. The damping values discussed above are equivalent viscous damping ratios, however the mechanism of damping is hysteretic (with possible additional contributions from impact effects) and depends on the area of lateral load—lateral displacement loops (Figure 6). Further study needs to be performed to determine if hysteretic energy dissipation is accounted for explicitly, what level of additional damping is provided by radiation impact damping.

ACKNOWLEDGEMENTS

Piletech Ltd. provided and installed the screw piles and also funded some of the site investigation work. The NZ Earthquake Commission (EQC) also contributed to the costs of the project. The contribution from the team of helpers, without which the fieldwork could not have been successfully completed, is gratefully acknowledged: Ryan Lefebve, Wesley Bell, Asaah Frimpong, Kiersten Bakke, Katie Eberle, Peter Kendicky, Sadeq Asadi, Baqer Asadi, Arman Kamalzadeh, Ravi Salimath, Marjus Gjata. Finally, the contributions of those who assisted with the construction of the field testing rigs is acknowledged: Jay Naidoo, Shane Smith, Andrew Virtue, Mark Byrami, Ross Reichardt, Jeff Melster, Mark Liew, and Mark Twiname.

REFERENCES

Anderson, J.B., F.C. Townsend and B. Grajales (2003). Case history evaluation of laterally loaded piles. *Journal of Geotechnical and Geoenvironmental Engineering* 129(3): 187–196.

Blaney, G.W. and M.W. O'Neill (1986). Measured lateral response of mass on single pile in clay. *Journal of Geotechnical Engineering* 112(4): 443–457.

Boominathan, A. and R. Ayothiraman (2006). Dynamic response of laterally loaded piles in clay. *Proceedings of the Institution of Civil Engineers—Geotechnical Engineering* 159(3): 233–241.

Chopra, A.K. (2007). *Dynamics of Structures: Theory and Applications to Earthquake Engineering.* Upper Saddle River, New Jersey, Pearson Prentice Hall.

Cox, W., L.C. Reece and B.R. Grubbs (1974). Field testing of laterally loaded piles in sand. *6th Offshore Technology Conference.* Houston: 459–472.

Edbrooke, S.W. (2001). *Geology of the Auckland Area*, Institute of Geological and Nuclear Sciences: 1:250 000 geological map 253.

Inman, D.J. (2014). *Engineerig Vibration* 4th edition. Harlow, Essex, England, Pearson Education.

Janoyan, K.D., J.W. Wallace and J.P. Stewart (2006). Full-scale cyclic lateral load test of reinforced concrete pier-column. *ACI Structural Journal* 103(2): 178–187.

Reece, L.C. and R.C. Welch (1975). Lateral loading of deep foundations in stiff clay. *Journal of Geotechnical Engineering* 101: 633–649.

Robertson, P.K. and K.L. Cabal (2010). *Guide to Cone Penetration Testing for Geotechincal Engineering.*

Sa'don, N.M., M.J. Pender, A.R. Abdul Karim and R. Orense (2014). Pile Head Cyclic Lateral Loading of Single Pile. *Geotechnical and Geological Engineering* 32(4): 1053–1064.

Seismic Performance of Soil-Foundation-Structure Systems – Chouw, Orense & Larkin (Eds)
© 2017 Taylor & Francis Group, London, ISBN 978-1-138-06251-1

Investigation of the influence of soil-structure interaction on the seismic performance of a skewed bridge

C. Kun & N. Chouw
Department of Civil and Environmental Engineering, The University of Auckland, Auckland, New Zealand

L. Jiang
National Engineering Laboratory for Construction Technology of High Speed Railway, The Central South University, China

ABSTRACT: Skewed bridges are known to be more susceptible to girder unseating than straight bridges. This increased vulnerability has been attributed to the out-of-plane movements of the girders. One of the major factors that affect the seismic response of skewed bridges is the Soil-Structure Interaction (SSI). The effects of SSI and skew angle applied simultaneously have not been explored in the past. In this investigation, 1:20 scale straight and skewed bridges were subjected to seismic excitations. The ground motion was simulated based on the NZS 1170.5 design spectrum for shallow soil condition (Class C). The bridges were fixed on a shake table to simulate the idealized fixed base case and placed on compacted sand to include the influence of SSI. The effects of SSI on the girder and footing displacements, and rotation of the girder were discussed. The findings provide useful insight into the response of skewed bridges including the effects of SSI.

1 INTRODUCTION

Skewed bridges have been known to be more prone to damage during earthquakes than straight ones. The increased vulnerability of skewed bridges has been associated with the rotations of the girders, causing large out-of-plane displacements. The catastrophic effects of these displacements have been seen in many earthquakes in the past for example in the 1971 San Fernando earthquake (Wood & Jennings 1971) and the 1995 Kobe earthquake (Chouw 1995).

Although studies on the seismic response of skewed bridges have been quite extensive over the past few decades, most of them have did not consider a number of factors that could influence the responses. Many of them such as by Maragakis (1985) and Meng *et al.* (2001) have considered the response of the skewed bridges under only ideal fixed base conditions. The significance of soil-structure interaction (SSI) on the response of skewed bridges has only been studied in recent years e.g. by Mallick and Raychowdhury (2015).

Conventionally, SSI is assumed to be beneficial in reducing the response of bridges due to the flexibility of the underlying soil. Ciampoli & Pinto (1995) found that although SSI increased the maximum displacement of the piers, the inelastic curvature demand on the piers remain unaffected, and in fact tends to decrease. However, in recent years, many studies have found the opposite to be true, and that the conventional assumption was oversimplified. Mylonakis & Gazetas (2000) found that SSI could increase the ductility demands of inelastic piers. Chouw (2008) also found that design specifications based on the assumption of fixed base conditions could severely underestimate the required seat length of bridge girders.

To the authors' best knowledge, the only study on skewed bridges that includes the effects of the underlying soil foundation was a numerical investigation carried out by Ghotbi (2014). It was found that the effects of SSI depend on the foundation soil, but typically, the softer the soil, the higher the chances of damage to the bridge. No experimental work has been conducted on the topic to date. This study aims to provide useful insight into the influence of SSI on the seismic response of a skewed bridge through shake table testing. Comparison of the results from the idealized fixed base case and when the bridge is founded upon soil will be made.

2 METHODOLOGY

2.1 *Bridge prototype and model*

The Newmarket Viaduct Replacement Bridge located in Auckland, New Zealand was chosen as the prototype. The length of the bridge girder was 100 m, with a pier-to-pier distance of 50 m and height of 15.5 m. The fundamental frequency in the longitudinal direction was calculated to be 0.98 Hz. The prototype was scaled based on principles of similitude outlined by Dove and Bennett (1986) to obtain dimensions for large-scale bridge model. A length scale factor of 20 was applied to the prototype. The fundamental frequency of the bridge model was maintained same as that of the prototype. The footing was also scaled accordingly and the size obtained was 500 mm by 300 mm. A 30° skewed bridge with similar dimensions as that of the straight was constructed.

In the idealized fixed base case, the straight and skewed bridges and the abutments were bolted onto the shake table. In the SSI case, sand was placed in a rigid box with inner dimensions 3.5 m by 1.5 m by 2 m, on which the bridge segment was placed. 8 m³ of loose, wet sand was placed in the box and compacted to about 1.5 m of depth. Some of the parameters of the soil used are shown in Table 2.

Table 1. Dimensions of prototype and model.

Parameter	Prototype	Model
Length of girder	100 m	5 m
Pier height	15.5 m	775 mm
Distance between piers	50 m	2.5 m
Pier width	3.44 m	100 mm
Pier thickness	1.48 m	6 mm
Footing width	10 m	500 mm
Footing length	6 m	300 mm
Fundamental frequency	0.98 Hz	0.98 Hz

Table 2. Parameters of soil.

Parameter	
Water content, w	24%
Specific gravity, G	2.70
Minimum dry density, $\rho_{d, min}$	1450 kg/m³
Maximum dry density, $\rho_{d, max}$	1810 kg/m³

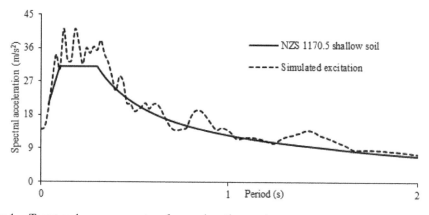

Figure 1. Target and response spectra of ground motion used.

2.2 *Ground motion*

The ground motion used was stochastically simulated based on the target spectrum specified in the NZS 1170.5 for shallow soil (Class C) conditions. The design spectrum and response spectrum of the simulated ground motion are shown in Figure 1.

3 RESULTS AND DISCUSSION

The displacements of the bridge girder, in both the fixed base and SSI cases, and of the footings, in the case of SSI, were measured using linear variable differential transformers (LVDTs) and laser transducers. Figure 2 shows the displacements of the girders of the straight and skewed bridges in the longitudinal direction. It can be seen that SSI has different effects on the bridge responses depending on the type of bridge. For the straight bridge, a relatively small reduction in the maximum displacement from 13.77 mm to 11.37 mm was seen when SSI was introduced. However, for the 30° skewed bridge, SSI seemed to greatly increase the maximum displacement from 5.75 mm to 8.90 mm.

The vertical displacement of the footing of each bridge was plotted in Figure 3. The straight bridge had a maximum footing displacement of about 0.12 mm, whereas that of the skewed bridge was about 0.18 mm, a 1.5 times increase.

The larger footing displacement of the skewed bridge most likely induced larger torsional responses of the girder, hence increasing the displacement demands. The rotation of the girder in both the fixed base and SSI cases were shown in Figure 4. The maximum rotation of

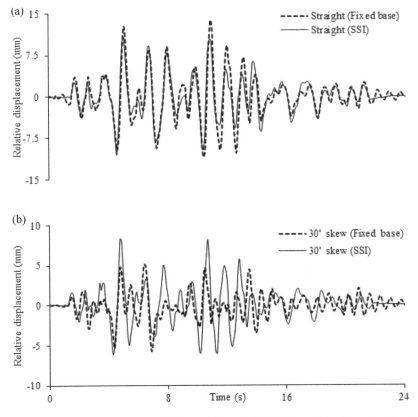

Figure 2. Longitudinal displacement of (a) straight and (b) 30° skewed bridge under the idealized fixed base and SSI conditions.

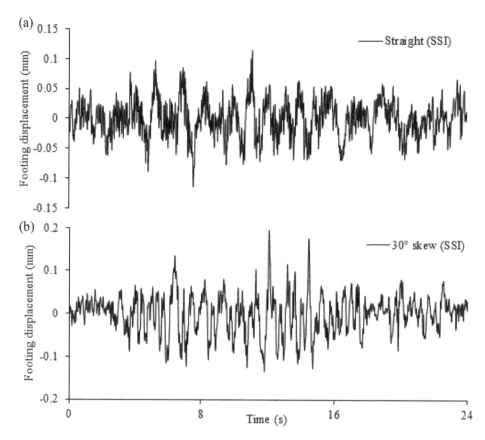

Figure 3.   Footing displacement of (a) straight and (b) 30° skewed bridge for the SSI case.

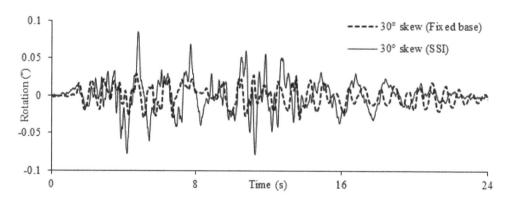

Figure 4.   Rotation of girder of 30° skewed bridge.

the girder in the idealized fixed base case was 0.036°, whereas when the bridge was founded on sand, the maximum rotation was significantly increased to 0.085°.

4   CONCLUSIONS

This paper discusses the influence of SSI on the seismic response of a straight and skewed bridge through experimental studies. Comparison was made between two cases: Idealized fixed base case, where the bridges were fixed on the shake table, and the SSI case, where the bridges were founded upon compacted sand placed in a rigid sand box. The following conclusions are drawn:

- SSI reduces the girder displacement of the straight bridge in the longitudinal direction, but causes a significant increase for the 30° skewed bridge.
- A possible reason for the increase in girder displacement of the skewed bridge when SSI was introduced was the larger footing displacement compared to that of the straight bridge.
- The larger footing displacement most likely induced larger torsional response, and consequently, larger rotation of the girder of the skewed bridge.
- The larger rotation could be significant when designing for the out-of-plane movements of skewed bridges.

ACKNOWLEDGEMENTS

The authors would like to acknowledge the financial support from the China Railway Corporation, project number 2013G002-A-1. The authors would also like to thank the Ministry of Business, Innovation and Employment for the support through the Natural Hazards Research Platform under the Award 3708936.

REFERENCES

Chouw, N. 1995. Effect of the earthquake on 17th of January 1995 on Kobe. *Proceedings of the D-A-CH meeting of the German, Austrian and Swiss Society for Earthquake Engineering and Structural Dynamics, University of Graz, Austria*: 135–169.

Chouw, N. 2008. Unequal soil-structure interaction effect on seismic response of adjacent structure. *Proceedings of the 18th New Zealand Geotechnical Society Symposium on Soil-structure Interaction—From rules of thumb to reality, University of Auckland, 4–5 September*: 214–219.

Ciampoli, M. & Pinto, P. 1995. Effects of soil-structure interaction on inelastic seismic response of bridge piers. ASCE *Journal of Structural Engineering* 121(5): 806–814.

Dove, R.C. & Bennett, J.G. 1986. Scale modeling of reinforced concrete Category I structures subjected to seismic loading. *Los Alamos National Lab., NM (USA)*. Retrieved from http://www.osti.gov/scitech/servlets/purl/6084387.

Ghotbi, A.R. 2014. Performance-based seismic assessment of skewed bridges with and without considering soil-foundation interaction effects for various site classes. *Earthquake Engineering and Engineering Vibration* 13(3):357–373.

Mallick, M. & Raychowdhury, P. 2015. Seismic analysis of highway skew bridges with nonlinear soil-pile interaction. *Transportation Geotechnics* 3: 36–47.

Maragakis, E. 1985. A model for the rigid body motions of skew bridges. *Casadena, California: California Institute of Technology*. Retrieved from http://www.scopus.com/inward/record.url?eid = 2-s2.0-0021938781&partnerID = 40&md5 = 213fecabcd83a8e2c7e5c23349bb4ab2.

Meng, J.Y., Lui, E.M. & Liu, Y. 2001. Dynamic response of skew highway bridges. *Journal of Earthquake Engineering* 5(2): 205–223.

Mylonakis, G. & Gazetas, G. 2000. Seismic soil-structure interaction: beneficial or detrimental? *Journal of Earthquake Engineering* 4(3):277–301.

Wood, J. & Jennings, P. 1971. Damage to freeway structures in the San Fernando earthquake. *Bulletin NZ Society of Earthquake Engineers* 4(3): 347–376.

Seismic Performance of Soil-Foundation-Structure Systems – Chouw, Orense & Larkin (Eds)
© 2017 Taylor & Francis Group, London, ISBN 978-1-138-06251-1

Soil-foundation-structure-fluid interaction during earthquakes

S. Iai & K. Ueda
Disaster Prevention Research Institute, Kyoto University, Japan

T. Tobita
Department of Civil, Environmental and Applied System Engineering, Kansai University, Japan

ABSTRACT: In this study, a centrifuge model tests and effective stress analyses are performed on a steel frame building with pile-foundation subject to Tsunami such as those seriously damaged during the 2011 off the Pacific coast of Tohoku Earthquake (Magnitude 9.0). The centrifuge model tests at a scale of 1/200 in geometry are performed to simulate the overturning of a building subject to combined effects of soil liquefaction and Tsunami. With the effective stress analyses, this study demonstrates the importance of the combined effects of soil liquefaction at pile foundations in addition to the external force of Tsunami.

1 INTRODUCTION

An earthquake with a Japan Meteorological Agency (JMA) magnitude 9.0 hit north east Japan at 14:46, March 11, 2011. JMA named this earthquake 'the 2011 off the Pacific coast of Tohoku Earthquake'. This earthquake is the greatest in its magnitude since the modern earthquake monitoring system was established in Japan.

Recorded heights of the Tsunamis were higher than 7.3 m at Soma, higher than 4.2 m at Oarai, and higher than 4.1 m at Kamaishi. The impact of the Tsunamis is also the greatest since the existing design methodology was adopted for designing structures constructed along coast lines.

The most typical example of the damage to buildings constructed along coast lines was the one at Onagawa city. In this example, the RC or steel frame buildings supported by pile foundations were washed away and toppled down due to Tsunami. These types of buildings were supposed to resist the external force due to Tsunami and expected to function as evacuation facilities. Consequently, the damage to these types of buildings, including the pull-out of pile foundation, was unexpected, posing a new engineering problem related to the effects of Tsunami.

Figure 1 shows a typical damage to the steel frame buildings at Onagawa city. The four story building had dimension of 8.4 m by 16 m wide and 12 m high, receiving Tsunami force toward the shortest dimension. This building was overturned together with the footing, and the piles were damaged at the pile top near the footing. At the right side of the photo, the pulled out piles are dragged down from the footing. After the Tsunami, this building was found washed away 20 m over a parking area in the vicinity of the original location of the building.

The mechanism of failure of this type of damage to buildings was speculated to be due not only to the Tsunami force, but also combined effects due to liquefaction of foundation ground. In order to investigate the primary mechanism of failure of this type, the results of a series of centrifuge model tests and effective stress analyses are presented in this paper.

2 CENTRIFUGE MODEL TESTS

2.1 *Generator for simulating Tsunami in centrifuge test*

The centrifuge at Disaster Prevention Research Institute, Kyoto University (effective radius 2.5 m) was used for the centrifuge model tests in this study. The equipment for simulating

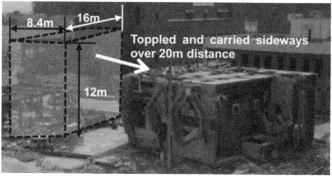

(a) Overview

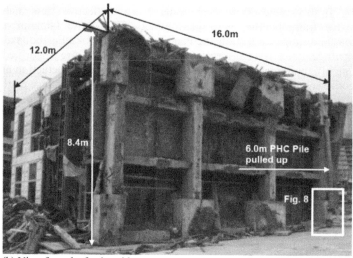

(b) View from the footing side

Figure 1.   Toppling of a four story steel frame building with pile-foundation at Onagawa city during 2011 East Japan earthquake (after JSCE, 2011).

Tsunami in the centrifuge is shown in Figure 2. In this equipment, the remote-control valve at the bottom of the water tank is opened to generate the Tsunami-like water flow toward a structure such as a building and overflowing water at the other end of the model (left in this figure) is absorbed in a tentative storage pit for reducing the effect of reflecting wave from the (left side) wall of the container.

The centrifuge model tests of a building with pile foundations were performed for a building 12 m high, 8.4 m by 16 m wide, in prototype with a scaling factor of 1/200 in geometry by adopting the generalized scaling relation in a 20 g centrifugal acceleration field. The cross section of the model is shown in Figure 3. In this model, piles were made of solid stainless steel rods, 1.2 m in diameter with a density of 7.93 t/m³. Pile tops were fixed to the base of the rigid model building but pile ends were placed on a rigid base made of a metal block, allowing separation from the rigid base when the piles are pulled out.

Silica No. 7 sand with a relative density of 50% was used for the liquefiable sand deposit. Figure 4 shows the liquefaction resistance curve of this sand obtained from a series of cyclic triaxial tests. The thickness of this sand deposit was varied from 4.6 m to 16.6 m. Pile length was adjusted to the same as the thickness of the sand deposit. Viscous fluid was used following the generalized scaling relationship. Pure water was used for Tsunami wave because the effect of seepage flow from the ground surface is considered negligibly small.

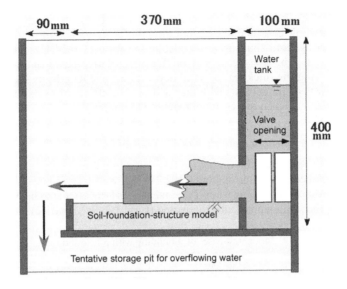

Figure 2. Equipment for simulating Tsunami in centrifuge model tests and model of a breakwater (unit in mm in model scale).

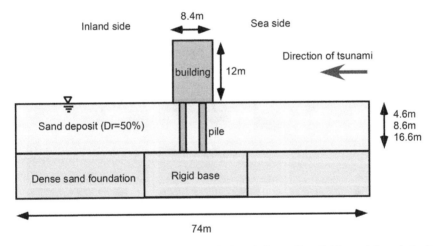

Figure 3. Centrifuge model of a building with pile foundation at liquefiable sand deposit (unit in m in prototype scale).

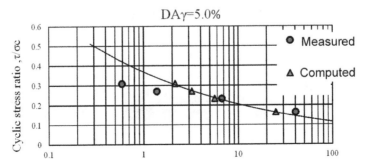

Figure 4. Liquefaction resistance curve of Silica No.7 sand.

Building was made of a mortar block with a unit weight adjusted to the bulk density of RC building. The whole equipment for simulating the Tsunami was mounted on a shaking table in the centrifuge and shaken by a sinusoidal motion of 0.54 Hz, with a maximum acceleration of 0.25 g before the arrival of Tsunami. Three cases of model tests were performed by varying the thickness of sand deposit and pile lengths as shown in Table 1.

Each case consists of model tests with and without the effects of excess pore water pressure in the sand deposit when the Tsunami reaches the building.

2.2 *Centrifuge test results of a building with pile foundation*

Figure 5 shows the results of the centrifuge tests of a building with pile foundation when Tsunami arrives immediately after shaking (Case 1-1, 2-1, and 3-1). Figure 6 shows the results of the centrifuge tests of a building when Tsunami arrives long after shaking with excess pore water pressure in the sand deposit fully dissipated (Case 1-2, 2-2, and 3-2).

Table 1. Model test cases of a building with pile foundation.

Case	Pile length m	Excess pore pressure (at the time of Tsunami)
Case 1-1	4.6	remaining
Case 1-2	4.6	fully dissipated
Case 2-1	8.6	remaining
Case 2-2	8.6	fully dissipated
Case 3-1	16.6	remaining
Case 3-2	16.6	fully dissipated

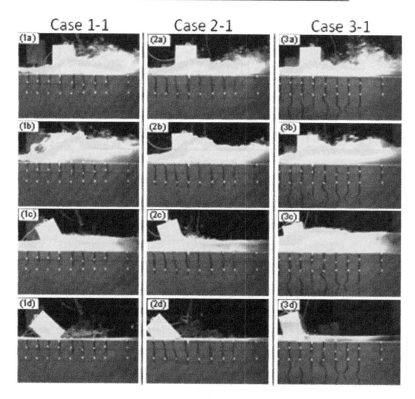

Figure 5. Deformation of a building with pile foundation when Tsunami arrives immediately after shaking (with remaining excess pore water pressures in the liquefiable sand deposit): Cases 1-1, 2-1, and 3-1.

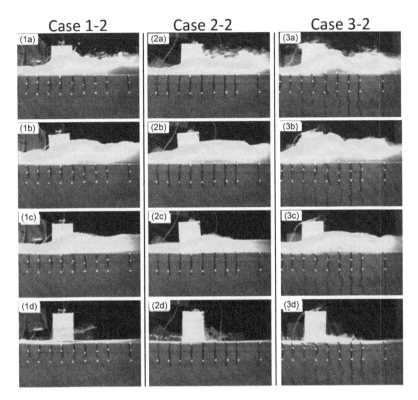

Figure 6. Deformation of a building with pile foundation when the Tsunami arrives long after shaking (with fully dissipated excess pore water pressures in the liquefiable sand deposit): Cases 1-2, 2-2, and 3-2.

As shown in Figures 5 (1b), (2b), and (3b), when the Tsunami arrives and the Tsunami force is applied to the building, the building begins to tilt about 5 degrees toward the downstream side of the Tsunami. This degree of deformation is commonly observed in Figure 6 (1b), (2b), and (3b). However, difference between the results shown Figures 5 and 6 becomes distinct in (1c), (2c), and (3c) or (1d), (2d), and (3d).

The tilting of the building continues to progress when the Tsunami arrives immediately after the shaking whereas the building has returned to the original straight position when the Tsunami arrives long after the shaking. Thus, the centrifuge model tests of a building successfully demonstrated that combined effects of liquefaction of ground with Tsunami force can be the major mechanism of failure of a building with pile foundations.

The model test also demonstrated that the degree of tilting is larger for a building supported by shorter piles. This trend should be considered as one of the important factors when studying mitigation measures against the possible failure of a building with pile foundation in liquefiable ground.

3 EFFECTIVE STRESS ANALYSIS

3.1 Effective stress model

For the effective stress analysis of a building subject to combined effects of soil liquefaction and wave action, the strain space multiple mechanism model with a cocktail glass model is used (Iai et al. 2011). In the strain space multiple mechanism model, the effective stress, defined as extension positive, is given based on a dyad defined by the unit vector $\mathbf{n}$ along the direction of the branch between the particles in contact with each other and the unit vector $\mathbf{t}$ normal to $\mathbf{n}$ as follows:

85

$$\sigma' = -p\mathbf{I} + \frac{1}{4\pi}\iint q\langle \mathbf{t}\otimes\mathbf{n}\rangle \mathrm{d}\omega\mathrm{d}\Omega \tag{1}$$

$$\langle \mathbf{t}\otimes\mathbf{n}\rangle = \mathbf{t}\otimes\mathbf{n} + \mathbf{n}\otimes\mathbf{t} \tag{2}$$

where p denotes effective confining pressure (compression positive), $\mathbf{I}$ denotes second order identity tensor, q denotes micromechanical stress contributions to macroscopic deviator stress due to virtual simple shear mechanism (called virtual simple shear stress), and $\langle \mathbf{t}\otimes\mathbf{n}\rangle$ denotes second order tensor representing the virtual simple shear mechanism. In the double integration, the integration with respect to $\omega (= 0$ through $\pi)$ is taken over a virtual plane spanned by the direction vectors $\mathbf{n}$ and $\mathbf{t}$ with $\omega/2$ being the angle of $\mathbf{n}$ relative to the reference local coordinate defined in the virtual plane, while the integration with respect to the solid angle Ω is taken over a surface of a unit sphere to give a three dimensional average of two dimensional mechanisms.

The integrated form of the constitutive equation, i.e. direct stress strain relationship, is derived by relating the macroscopic strain tensor $\boldsymbol{\varepsilon}$ to the macroscopic effective stress σ' through the structure defined by Equation (1). The first step to derive this relationship is to define the volumetric strain ε (extension positive) and the virtual simple shear strains γ as the projections of the macroscopic strain field to the second order tensors representing volumetric and virtual simple shear mechanisms as follows:

$$\varepsilon = \mathbf{I}:\boldsymbol{\varepsilon} \tag{3}$$

$$\gamma = \langle\mathbf{t}\otimes\mathbf{n}\rangle:\boldsymbol{\varepsilon} \tag{4}$$

where the double dot symbol denotes double contraction. In order to take into account the effect of volumetric strain due to dilatancy ε_d, effective volumetric strain ε' is introduced by

$$\varepsilon' = \varepsilon - \varepsilon_\mathrm{d} \tag{5}$$

where the rate of volumetric strain due to dilatancy is given by the projection of strain rate field to a second order tensor $\mathbf{I}_\mathrm{d}$ as

$$\dot{\varepsilon}_\mathrm{d} = \mathbf{I}_\mathrm{d}:\dot{\boldsymbol{\varepsilon}} \tag{6}$$

The scalar variables defined in Equations (4) and (5) as the projection of macroscopic strain field are used to define the isotropic stress p and virtual simple shear stress q in Equation (1) through path dependent functions as

$$p = p(\varepsilon') \tag{7}$$

$$q = q(\gamma) \tag{8}$$

In the strain space multiple mechanism model, the virtual simple shear mechanism is formulated as a non-linear hysteretic function, where a back-bone curve is given by the following hyperbolic function;

$$q(\gamma) = \frac{\gamma/\gamma_\mathrm{v}}{1+|\gamma/\gamma_\mathrm{v}|}q_\mathrm{v} \tag{9}$$

The parameters q_v and γ_v defining the hyperbolic function are the shear strength and the reference strain of the virtual simple shear mechanism, respectively.

The isotropic component in Equation (7) is defined by a hysteretic tangential bulk modulus depending on the loading/unloading (L/U) condition as

$$K_\mathrm{L/U} = -\frac{\mathrm{d}p}{\mathrm{d}\varepsilon'} = r_K K_\mathrm{U0}\left(\frac{p}{p_0}\right)^{l_K} \tag{10}$$

where p_0: initial confining pressure, K_{U0}: tangential bulk modulus at initial confining pressure.

Dilatancy in Equation (5) in the Cocktail glass model is decomposed into contractive component ε_d^c and dilative component ε_d^d as

$$\varepsilon_d = \varepsilon_d^c + \varepsilon_d^d \tag{11}$$

In this study, dilatancy of the liquefiable sand deposit was idealized based on the liquefaction resistance shown in Figure 4. Using the model parameters of dilatancy calibrated to the liquefaction resistance, computed results are also shown in Figure 4. The effect of dilatancy in the dense foundation ground below the liquefiable sand deposit and the rubble mound for a breakwater model was assumed negligibly small and dilatancy was ignored.

3.2 Load conditions for simulating Tsunami and effect of liquefaction acting on a building with pile foundation

The effective stress analysis of a building with pile foundation was performed for a prototype scaled from the centrifuge model, including the building 12 m high. As shown in Figure 7, joint element is specified at the bottom of the building to allow sliding and separation between the building footing and foundation ground. Joint element is also specified for allowing sliding between the piles and the foundation ground. After static gravity analysis to set the initial conditions of building-foundation system, a sinusoidal input motion for a duration of 20 seconds, shown in Figure 8, was applied at the base of the foundation soil, and then after certain period for allowing dissipation of excess pore water pressure from the foundation ground, a Tsunami wave force was applied on the building as equivalent static distributed force as shown in Figure 9.

For simplicity, the analysis was performed with (Case 1) and without (Case 2) shaking before arrival of Tsunami. In Case 1, ten seconds were allowed between the end of shaking and arrival of Tsunami. Increasing rate of Tsunami height was simulated for allowing 40 seconds from the arrival of Tsunami to the maximum Tsunami height of 6.0 m.

3.3 Results of the effective stress analysis of a building with pile foundation

The results of analysis of Case-2-1 & 2-2 of a building with pile foundation are presented below. Figure 10 (a) shows residual deformation of a breakwater with both Tsunami wave force on the building with foundation ground at the state of liquefaction. The tilting of the building is associated with a significant deformation of pile foundation and foundation ground. This mode of failure is consistent with that observed at the centrifuge model test. In comparison to this result, deformation of a building due to Tsunami wave force only is small as shown in Figure 10 (b). These results of the analyses indicate that primary mechanism of failure of a building with pile foundation due to Tsunami is combined failure mechanism due to Tsunami wave force and liquefaction of foundation ground.

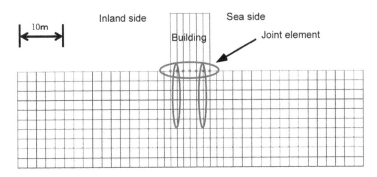

Figure 7. Finite element mesh of a building for analysis (prototype scale) and load conditions for simulating Tsunami wave force.

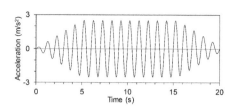

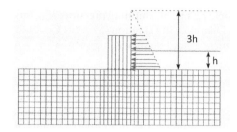

Figure 8.   Input motion for a building with pile foundation.

Figure 9.   Load conditions for simulating wave force due to Tsunami.

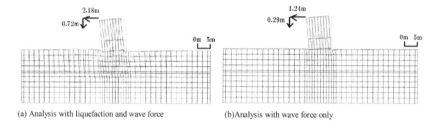

(a) Analysis with liquefaction and wave force   (b)Analysis with wave force only

Figure 10.   Residual deformation of a building with pile foundation.

The suction effects of the Tsunami is not considered in the analysis. If these effects are added in the analysis, induced deformation and tilting of the building can be larger than those presented in this analysis. However, the difference in the deformation between the analysis with both liquefaction and wave force and that with wave force only may remain about the same degree of orders presented in this paper.

4   CONCLUSIONS

In this study, a centrifuge model tests and effective stress analyses are performed on a breakwater and a building with pile foundation subject to a tsunami such as those which caused serious damaged during 2011 East Japan Earthquake (Magnitude 9.0). For a building with pile foundations, both the centrifuge model tests at a scale of 1/200 and the effective stress analyses demonstrate the importance of the combined mechanism of failure with the effect of liquefaction of foundation ground in addition to the wave force produced by tsunami action.

As a whole, this study demonstrates the importance of combined failure mechanism associated with the action of water on soil-structure systems.

REFERENCES

Iai, S., Tobita, T. & Nakahara, T. 2005. Generalized scaling relations for dynamic centrifuge tests, *Geotechnique*, 55(5): 355–362.

Iai, S., Tobita, T., Ozutsumi, O. & Ueda, K. 2011. Dilatancy of granular materials in a strain space multiple mechanism model, *International Journal for Numerical and Analytical Methods in Geomechanics*, 35(3): 360–392.

JSCE (Earthquake Engineering Committee) 2011. 2011 East Japan earthquake reconnaissance report, JSCE, No. 20 (in Japanese).

Seismic Performance of Soil-Foundation-Structure Systems – Chouw, Orense & Larkin (Eds)
© 2017 Taylor & Francis Group, London, ISBN 978-1-138-06251-1

A mathematical approach to computing structural-failure boundaries

H.M. Osinga
Department of Mathematics, University of Auckland, New Zealand

ABSTRACT: Earthquakes can cause substantial damage to buildings in ways that are still not well understood. The amplitude and principal frequency of an earthquake are two primary components that affect the extent of the damage, and they are the basis for many design specification guidelines. We investigate how an external forcing with varying amplitude and principal frequency affects structurural stability. As an example we consider a model of a planar, post-tensioned frame that exhibits dynamics quite similar to the experimental measurements of a scaled model on a shake table. Our goal is to predict behaviour of models subject to an aperiodic external forcing (an earthquake). Here, we consider a periodic external forcing, which is a simplifying but common choice. Many results in the literature are obtained from performing a large number of simulations over a range of amplitudes and frequencies. Our approach is much more efficient and uses a novel computational method that approximates the failure boundary directly. We find that failure can occur in profoundly different ways, due to inherent nonlinearities in the system. Stability is particularly affected if the natural frequency of the structure is close to that of the external forcing.

1 INTRODUCTION

There have been quite a number of earthquakes recently, including in New Zealand, that raised the awareness of a need for earthquake resistant buildings. Most notorious is the 2011 Christchurch earthquake, which followed a series of earthquakes starting in September 2010 and provides a striking example of the need for better damage assessment (Kam et al. 2011). First and foremost, we like to have buildings that do not collapse during an earthquake so that lives are saved; even better, the building should be such that it sustains virtually no damage from any earthquake below a critical magnitude, so that costly reparations can be avoided and it can safely be used again after the event. Ideas for low-damage design include allowing a degree of damage at predefined locations that do not affect the safety of inhabitants (Priestley et al. 1999, Qin et al. 2013), and activating rigid body movement of structural members so that forces related to local deformation in the structure will be prevented (Acikgoz & De Jong 2012, Alexander et al. 2011, Fardis & Rakicevic 2012). Furthermore, mathematical models are developed that complement the experimental results with detailed numerical analysis (Alexander et al. 2011, Oddbjornsson et al. 2012).

A major drawback of the theoretical research is the fact that the earthquake is typically modelled by a sine wave, which effectively means that the theoretical results underestimate the resilience of the model. The main argument against using more complicated external forcing terms is the simplicity of reducing the system to an autonomous equation. The response of such systems is governed by periodic solutions that have the same period as the sinusoidal earthquake, which implies that the analysis can be done with standard software packages. We explore a different approach that offers the possibility of computing failure boundaries of the model directly, without the need to formulate the system in autonomous form. As an example, we consider the model of a tied rocking block on an elastic foundation, which is equivalent to that of a planar, post-tensioned frame on a shake table (Alexander et al. 2011).

For such frames, the joints between beam-columns and column-foundations are held together by pre-stressed cables, and the elastic nonlinearity of the frame is entirely determined by the mechanics of these joint connections. The system dynamics can then be described in terms of a standardised (non-dimensional) angle φ that is equal to ± 1 at the point of joint opening. This leads to the equation

$$\begin{cases} \ddot{\varphi} + 2\gamma\dot{\varphi} + \mu(\varphi) = A\sin(\omega t), \\ |\varphi(t)| < \varphi_{\max} \quad \text{for all} \quad 0 \le t \le T_{\text{end}}, \end{cases} \tag{1}$$

where, the dot represents derivation with respect to time t. The maximum angle $\varphi_{\max}$ depends on the characteristics of the building and T_{end} is some maximum integration time that represents the duration of the earthquake. The stiffness function $\mu(\varphi)$ is given by

$$\mu(\varphi) = \begin{cases} \varphi, & |\varphi| \le 1, \\[2mm] \left(\dfrac{3}{\beta} + \dfrac{12}{\beta^2} + \dfrac{8}{\beta^3}\right)\varphi \\[2mm] \quad + \left(3 + \dfrac{9}{\beta} + 6\dfrac{1-\sqrt{\psi}}{\beta^2} - 6\dfrac{\sqrt{\psi}}{\beta^3} - 2\dfrac{\psi\sqrt{\psi}}{\beta^3\phi^2}\right)\mathrm{sgn}(\varphi), & |\varphi| > 1, \end{cases}$$

where $\psi = (1+\beta)(\varphi^2 + \beta|\varphi|)$ and β is the contact-to-cable stiffness ratio; see (Alexander et al. 2011) for details. Solutions φ to (1) are called *admissible* if $|\varphi(t)| < \varphi_{\max}$ for all t; only admissible solutions to (1) are truthful representations of actual behaviour that can be reproduced in experiments. We use the same parameters as in (Alexander et al. 2011), that is, we fix $\beta = 85$, $\gamma = 0.05$ and $\varphi_{\max} = 10$, and we consider periodic ground motion with varying frequency ω and amplitude A. Here, ω and A are non-dimensional parameters; the frequency ω is the ratio between the actual forcing frequency and the natural frequency of the frame; and the amplitude A involves the peak ground acceleration relative to the scaled angle and the storey height of the building; see (Alexander et al. 2011).

We are particularly interested in the solution $\varphi(t)$ of (1) that satisfies the initial condition $(\varphi(0), \dot{\varphi}(0)) = (0,0)$; we denote this solution by $\Phi_0(t)$. Our goal is to understand how admissibility of $\Phi_0(t)$ depends on the forcing frequency ω and amplitude A. Since the forcing is periodic, any bounded solutions will eventually be periodic. Consequently, it seems reasonable to expect that it is possible to predict admissibility of $\Phi_0(t)$ from the admissibility of the limiting periodic orbit. We argue that such prediction is not possible, not even in an approximating sense.

2 FORCING WITH FIXED FREQUENCY $\omega = 0.575$

For a range of pairs (ω, A) in the frequency-amplitude plane, there exist three different periodic orbits and two of these are stable. Such bistability is well known to occur in nonlinear oscillator systems like system (1). Let us consider $\omega = 0.575$ fixed and consider the A-dependent family of periodic orbits, which can readily be computed, for example, via pseudo-arclength continuation with the software package AUTO (Doedel 2007).

2.1 *Admissibility of periodic orbits*

If the forcing amplitude $A = 0$ then the periodic orbit has zero amplitude and is, in fact, equal to $\Phi_0(t)$. Starting from this solution, we can compute a one-parameter family of periodic orbits by increasing A. Figure 1(a) shows the result of such a continuation with respect to A, where $\omega = 0.575$ is kept fixed at a value that is representative for a large range of forcing frequencies; here, we plot the maximum angle of the periodic orbit versus A. For small A, only one periodic orbit exists, which has low amplitude and is stable; we denote it by A_ℓ.

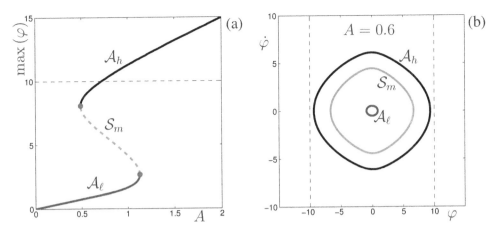

Figure 1. Periodic orbits of system (1) in dependence on the forcing amplitude A, where $\omega = 0.575$ is fixed. Panel (a) shows A on the horizontal axis and the amplitude of the periodic orbit on the vertical axis. Panel (b) shows three co-existing periodic orbits for $A = 0.6$ in the bistable regime plotted in projection onto the $(\varphi, \dot{\varphi})$-plane. The low- and high-amplitude periodic orbits labelled A_ℓ and A_h, respectively, are stable, while the mid-amplitude periodic orbit labelled S_m is of saddle type.

We checked that $|\Phi_0(t) < \varphi_{\max}|$ for all t when the forcing is chosen from the regime in the (ω, A)-plane for which only A_ℓ exists; hence, A_ℓ is always admissible. Provided A is small enough, $\Phi_0(t)$ accumulates onto A_ℓ.

As A increases, a fold bifurcation gives rise to a pair of periodic orbits, an attractor A_h and a saddle S_m, that have much larger amplitudes than A_ℓ. For $\omega = 0.575$, this fold bifurcation occurs at $A \approx 0.4897$. The fold bifurcation marks the beginning of a bistable regime during which the amplitude of the attractor A_ℓ increases and that of the saddle S_m decreases until A_ℓ and S_m merge and disappear at a second fold bifurcation; for $\omega = 0.575$, the second fold bifurcation occurs at $A \approx 1.1282$. For large A, only the attractor A_h exists and $\Phi_0(t)$ accumulates onto A_h. As indicated in Figure 1(a), the maximum of φ along A_h exceeds $\varphi_{\max}$ from $A \approx 0.7151$. Indeed, also for other forcing frequencies, the high-amplitude attracting periodic orbit A_h is not admissible for large values of A. Therefore, $\Phi_0(t)$ is not admissible for large A either.

Figure 1(b) shows the co-existence of three periodic orbits for the parameter pair $(\omega, A) = (0.575, 0.6)$ in projection onto the $(\varphi, \dot{\varphi})$-plane. Observe that A_ℓ has low amplitude and is clearly admissible, while the amplitude for A_h is so high that it is almost equal to $\varphi_{\max}$. Note that the maximum angle of these periodic orbits is exactly the same as their minimum angle, because of the symmetry $(\varphi, \dot{\varphi}, t) \mapsto (-\varphi, -\dot{\varphi}, t + \pi/\omega)$ of system (1). For $(\omega, A) = (0.575, 0.6)$, the solution $\Phi_0(t)$ is admissible and it accumulates onto A_ℓ.

2.2 Unexpected failure

Let us now consider the parameter pair $(\omega, A) = (0.575, 1.0607)$ for which there are still three co-existing periodic orbits, A_ℓ, S_m and A_h. For this higher value of A, the periodic orbits A_ℓ and A_h are both stable, but A_h is no longer admissible. Figure 2 shows the time series of $\Phi_0(t)$ overlayed on A_ℓ in panel (a) and on S_m in panel (b), with the corresponding projections onto the $(\varphi, \dot{\varphi})$-plane, in panels (c) and (d), respectively; included inpanel (d) is the periodic orbit A_h, which is not admissible. Observe from panel (a) that $\Phi_0(t)$ eventually accumulates onto A_ℓ, while panel (b) illustrates how $\Phi_0(t)$ first appears to accumulate onto S_m. Panels (c) and (d) clearly show that $|\Phi_0(t)|$ is much smaller that $\varphi_{\max}$ for all t.

Even though Figure 2 provides no indication of imminent danger, the amplitude $A = 1.0607$ is very close to the failure boundary for $\omega = 0.575$. Indeed, $\Phi_0(t)$ is not admissible when the amplitude is only slightly larger. Figure 3 shows the behaviour of $\Phi_0(t)$ for the parameter pair $(\omega, A) = (0.575, 1.0610)$. The figure is similar to Figure 2, showing the time series of $\Phi_0(t)$ in panels (a) and (b) and the corresponding phase portraits in the $(\varphi, \dot{\varphi})$-plane in panels (c)

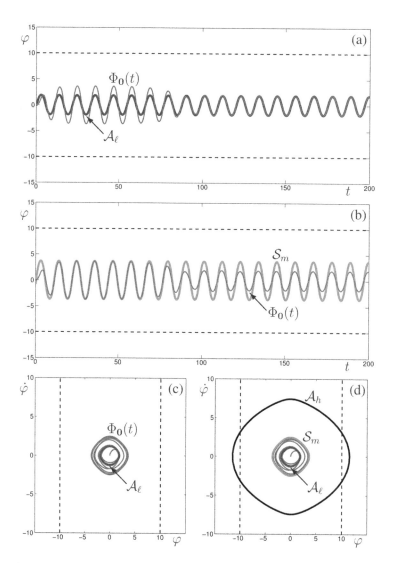

Figure 2. The solution $\Phi_0(t)$ of system (1) with $(\omega, A) = (0.575, 1.0607)$. Panel (a) compares the time series of $\Phi_0(t)$ with that of A_ℓ and panel (b) with that of S_m. The corresponding projections in the $(\varphi, \dot{\varphi})$-plane are shown in panels (c) and (d), respectively.

and (d), respectively. As in Figure 2, the solution $\Phi_0(t)$ again appears to accumulate onto S_m initially, but then, instead of decreasing in amplitude towards A_ℓ, the amplitude of $\Phi_0(t)$ increases and $\Phi_0(t)$ accumulates onto A_h. Since A_h is not admissible, admissibility of $\Phi_0(t)$ is also lost at this A-value. We note that failure of $\Phi_0(t)$ only occurs after a relatively long time in the admissible regime, which means that serious damage is only sustained if the duration of the forcing at this frequency and amplitude is long enough.

Observe that the transition from accumulation onto A_ℓ to accumulation onto A_h occurs well after A_h loses admissibility, at least for this value of ω. However, the maximum amplitude of $\Phi_0(t)$ is, in fact, larger than the amplitude of A_h, which indicates that failure of $\Phi_0(t)$ for other values of ω may occur even before admissibility of A_h is lost. We also note that the point of failure of $\Phi_0(t)$ for this forcing frequency occurs well before the fold bifurcation at $A \approx 1.1282$, where A_ℓ and S_m disappear. We estimate the precise A-value for $\omega = 0.575$ at

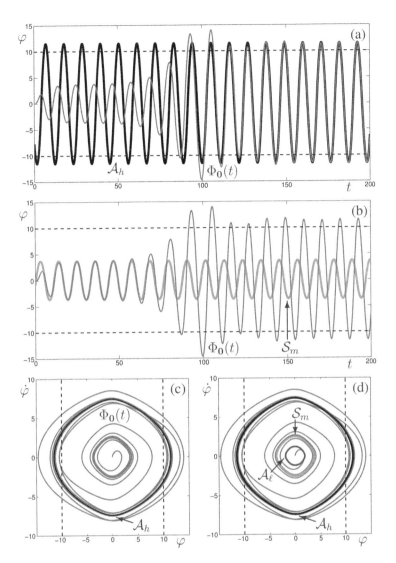

Figure 3. The solution $\Phi_0(t)$ of system (1) with $(\omega, A) = (0.575, 1.0610)$. Panel (a) compares the time series of $\Phi_0(t)$ with that of A_h and panel (b) with that of S_m. The corresponding projections in the $(\varphi, \dot{\varphi})$-plane are shown in panels (c) and (d), respectively.

which $\Phi_0(t)$ fails to be $A \approx 1.0608$. At this value, $\Phi_0(t)$ accumulates onto S_m instead of A_ℓ or A_h. Such behaviour is special, because S_m is not attracting. It means that $\Phi_0(t)$ is contained in the stable manifold of S_m, which is a surface in $(\varphi, \dot{\varphi}, t)$-space that separates the basins of attraction of A_ℓ and A_h. As illustrated in Figures 2 and 3, just before and just after loss of admissibility, respectively, neither the admissibility of the periodic orbits, nor the extent of the bistability regime can serve as a good approximation of the failure boundary. This means that the simplifying assumption of a periodic earthquake does not offer any benefit from an analytical point of view to help predict this type of failure.

3   FORCING WITH FIXED FREQUENCY $\omega = 0.675$

While the hard-to-predict failure described in the previous section occurs over a range of forcing frequencies, not all forcing frequencies lead to failure of this type. There also exists a range

93

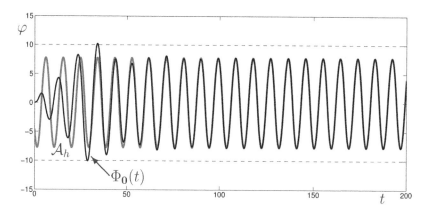

Figure 4. Time series of $\Phi_0(t)$ for $\omega = 0.675$ and $A = 0.95354140$, approximately at the moment of failure. The only existing periodic orbit at these parameter values, which is the high-amplitude periodic orbit A_h, is also shown.

of values for ω at which $\Phi_0(t)$ fails in a more predictable manner, namely, at which its amplitude increases in direct proportion with A until $|\Phi_0(t)| = \varphi_{max}$ for some $0 < t \leq T_{end}$. However, this type of failure can also not be predicted by studying the behaviour of the periodic orbits.

As an example, we consider $\omega = 0.675$, for which $\Phi_0(t)$ fails at $A \approx 0.9535$. At these values of ω and A, only one periodic orbit exists, A_h, which is admissible. Figure 4 shows the time series of $\Phi_0(t)$ at this (ω, A)-pair, overlaid on the time series of A_h. Even though A_h is admissible, the solution $\Phi_0(t)$ converges to it in a non-monotonic way and the angle φ of $\Phi_0(t)$ exceeds the amplitude of A_h during the transient approach to it. As shown in Figure 4, at about $t = 28.86$, the angle φ of $\Phi_0(t)$ grazes the boundary $\varphi = -10$, before $\Phi_0(t)$ converges to A_h. For slightly smaller A, the minimum of φ along $\Phi_0(t)$ remains just above $\varphi = -10$, while for slightly larger A, it will lie just below this lower bound.

While this type of failure is more gradual, it is important to realise that the periodic orbit A_h gives no indication of imminent failure, nor does the fact that there is no bistability for this value of A. The high-amplitude periodic orbit A_h only reaches the maximum amplitude of 10 when $A \approx 1.8043$ for $\omega = 0.675$. Furthermore, the low-amplitude periodic orbit A_ℓ for $\omega = 0.675$ exists up until $A \approx 0.8278$, at which it merges with the saddle periodic orbit S_m that appears in a fold bifurcation at $A \approx 0.3771$. Hence, also for this type of failure, there is no benefit, from an analytical point of view, in assuming that the earthquake is periodic.

4 FAILURE BOUNDARY IN THE (ω, A)-PLANE

We now consider admissibility of $\Phi_0(t)$ in dependence on both ω and A. Our approach is to approximate the moment of failure as the family of solutions $\Phi_0(t)$ that are tangent to the boundary of the admissible regime $\varphi(t) \in [-10, 10]$ for some $0 < t < T_{end}$; such a solution is called a *grazing solution*. Grazing solutions can be computed accurately, as is explained in the next section.

There are two types of grazing solutions: either $\Phi_0(t)$ is first tangent to the boundary $\varphi = -10$, which we call a left-grazing event, or $\Phi_0(t)$ is first tangent to the boundary $\varphi = +10$, which we call a right-grazing event. For any given frequency ω, there is typically more than one amplitude A for which $\Phi_0(t)$ is a grazing solution, and the failure boundary is defined as the curve with minimal amplitudes $A = A(\omega)$ for which there are grazing events. Figure 5 shows all curves of left- and right-grazing events for $\Phi_0(t)$ in the range $(\omega, A) \in [0.1, 1] \times [0.8, 2.6]$, up to total integration time $T_{end} = 150$; the darker-shaded curves, labelled g_L, correspond to left-grazing events and the lighter-shaded curves, labelled g_R, are right-grazing events. The choice $T_{end} = 150$ corresponds to about 10 periods of the forcing for $\omega = 0.575$, but substantially fewer or more periods when ω is close to 0.1 or 1, respectively.

94

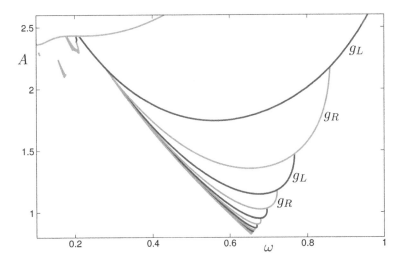

Figure 5. Left-grazing (g_L) and right-grazing (g_R) events for the solution $\Phi_0(t)$ of system (1) with $(\varphi, \dot{\varphi}) = (0,0)$ at $t = 0$.

There are no grazing events for $A < 0.8$. For larger A-values, a main resonance tongue can be discerned that approximately ranges over values $\omega \in [0.3,1]$. This main resonance tongue is composed of curves along which $\Phi_0(t)$ alternatingly grazes the left and right admissibility boundary. Note that the failure boundary is piecewise smooth, with a discontinuity in its slope at each point where a curve g_L meets a curve g_R. At such a double grazing event, $\Phi_0(t)$ is a left-grazing solution that also exhibits a right-grazing event at some later time $t < T_{\text{end}}$, or $\Phi_0(t)$ is a right-grazing solution that also exhibits a left-grazing event at some later time $t < T_{\text{end}}$. Hence, the failure boundary is not only piecewise smooth, there is also a discontinuity in the time at which grazing occurs when there is a double-grazing event. For example, if $\omega = 0.675$, the solution $\Phi_0(t)$ is admissible for all A below the first grazing event at $A \approx 0.9535$, at which $\Phi_0(t)$ exhibits a left-grazing event for $t \approx 28.86$. As ω increases, $\Phi_0(t)$ continues to graze the boundary $\varphi = -10$, where the time of grazing varies continuously with ω until $\omega \approx 0.6956$. For this value of ω, the solution $\Phi_0(t)$ grazes the boundary $\varphi = -10$ when $t \approx 28.64$, but also the boundary $\varphi = +10$ for $t \approx 23.55$, which is about half a period earlier. Hence, for slightly larger values of ω, the solution $\Phi_0(t)$ fails in a right-grazing event at a time approximately half a period earlier than before. Already in (Jennings & Husid 1968), it was reported that amplitude needed for a structure to fail after a relatively short duration is significantly higher than failure occurring after a long duration; Figure 5 indicates that the relationship between the duration of an earthquake and the necessary amplitude for a structure to fail is not continuous.

Note the accumulation of left- and right-grazing events for a range of values ω, including $\omega = 0.575$. Here, the failure boundary is characterised by the fact that $\Phi_0(t)$ accumulates on S_m instead of one of the two attractors A_ℓ or A_h, as described in Section 2.2.

Figure 5 also shows what could be interpreted as second and perhaps third harmonics that have not been fully resolved. Here, the value $T_{\text{end}} = 150$ represents only up to five periods of the forcing. It seems that other grazing events at higher periods form part of the failure boundary for these smaller ω-values, but a detailed investigation of this low-frequency regime is left for future work.

5 DIRECT COMPUTATIONAL METHOD

We now present an algorithm that computes the failure boundary directly as a curve in the (ω, A)-plane. Grazing events can be computed numerically by continuation of a two-point boundary value problem. To this end, we rewrite (1) as a system of first-order differential equations

$$\dot{\mathbf{u}} = T\,\mathbf{f}(\mathbf{u}), \tag{2}$$

where $\mathbf{u} = \{\mathbf{u}(s) := (\varphi(sT), \dot{\varphi}(sT), sT) \mid 0 \le s \le 1\}$ represents a trajectory or orbit segment of a solution φ of (1) up to time T. The orbit segment $\mathbf{u}$ is formulated in scaled time such that it is always defined on the interval $[0,1]$. We impose boundary conditions to ensure that the orbit segment starts with the particular initial condition $(\varphi, \dot{\varphi}) = (0,0)$ at time $t = 0$, and ends at a grazing point:

$$\begin{cases} \mathbf{u}(0) &= \quad (0,0,0), \\ \mathbf{u}(1) &= \quad (\pm 10, 0, 1). \end{cases} \tag{3}$$

With the integration time T and the frequency-amplitude pair (ω, A) as free parameters, system (2)–(3) is well posed and gives rise to one-parameter solution families that correspond to the left- and right-grazing events.

We use pseudo-arclength continuation with the software package AUTO (Doedel 2007) to find the left- and right-grazing solution families. To start the continuation, we first compute $\Phi_0(t)$ as the solution of system (1) starting from $\varphi = 0$ and $\dot{\varphi} = 0$ with $\omega = 0.575$ and $A = 5.0$; the high value of A ensures that $\Phi_0(t)$ crosses the admissibility boundaries $\varphi = \pm 10$ many times. We then use AUTO to find the values A below 5.0 at which $\Phi_0(t)$ exhibits a grazing event for $\omega = 0.575$ fixed, even when the event occurs after $\Phi_0(t)$ failed. Each of those grazing events can then be continued in ω and A to form a branch in the left- or right-grazing solution family.

We remark that the computation of grazing solutions is just as efficient if the sinusoidal forcing term is replaced by a different, more realistic wave form. Hence, it is straightforward to compute the failure boundary with respect to the principal frequency and maximal amplitude of a more complicated forcing, or a different choice of system parameters.

6 CONCLUSIONS

We showed that the response of a system subjected to a sinusoidal forcing depends highly nonlinearly on the wave frequency and amplitude. This dependence cannot be understood in terms of the periodic solutions of the system that exist as a result of the periodic forcing. Therefore, the simplification of a pure sine wave instead of a real earthquake does not lead to any benefit for a theoretical analysis. Instead, we proposed an efficient method that computes the failure boundary in the frequency-amplitude plane directly as the family of piecewise-smooth curves of grazing solutions. The accumulation of such curves marks the location of the parameter regime where unexpected failure occurs, in the sense that the system seems to be far from failure right up to reaching the failure boundary. This computational method is equally efficient for more realistic non-periodic forcings. We illustrated our findings with an example that models a planar, post-tensioned frame, but the results should be expected for any nonlinear model that exhibits bistability when experiencing a periodic forcing.

From a theoretical point of view, the question arises what type of wave form would be best to represent the worst-case scenario in a study of possible system behaviour. While the sine wave does not provide the expected computational benefits, it is generally understood to elicit a worst-case response. Other wave forms have been suggested and a thorough analysis has particularly focussed on wave forms that exhibit only a single or few oscillations, such as Ricker wavelets or other single- or several-cycle waves. Such wave forms have been studied, for example, in (Takewaki et al. 2012, Makris & Black 2004) and a detailed review using the so-called energy approach to modelling can be found in this volume (Takewaki & Kojima 2017). The approach taken here could be applied to pulse-type excitations, but is equally suitable for the study of system behaviour when subjected to other wavelet forms, such as those suggested in (Ryan 1994). One direction of further research is the interesting finding of self-similar behaviour reported in (Makris & Black 2004), which is also apparent in the main resonance tongue in Figure 5. A full analysis of system response when subjected to non-periodic wavelet-type forcings is left for future research.

ACKNOWLEDGEMENTS

The author is grateful to Chris Budd for helpful discussions on the geometry of the failure boundary. This research is supported by grant # 3712132 from the Engineering Faculty Research Development Fund of the University of Auckland.

REFERENCES

Acikgoz, S. & De Jong, M.J. 2012. The interaction of elasticity and rocking in flexible structures allowed to uplift. *Earthquake Engineering and Structural Dynamics* **41**(15): 2177–2194.

Alexander, N.A., Oddbjornsson, O., Taylor, C.A., Osinga, H.M. & Kelly, D.E. 2011. Exploring the dynamics of a class of post-tensioned, moment resisting frames. *Journal of Sound and Vibration* **330**(15): 3710–3728.

Doedel, E.J. 2007. AUTO-07P: Continuation and bifurcation software for ordinary differential equations. With major contributions from Champneys, A.R., Fairgrieve, T.F., Kuznetsov, Yu.A., Oldeman, B.E., Paffenroth, R.C., Sandstede, B., Wang, X.J. & Zhang, C.; available at http://cmvl.cs.concordia.ca/auto.

Fardis, M.N. & Rakicevic, Z.T. 2012. *Role of Seismic Testing Facilities in Performance-Based Earthquake Engineering: SERIES Workshop*. Heidelberg: Springer-Verlag.

Jennings, P.C. & Husid, R. 1968. Collapse of yielding structures during earthquakes. *Journal of Engineering Mechanics* **94**(5), 1045–1065.

Kam, W.Y., Pampanin, S. & Elwood, K. 2011. Seismic performance of reinforced concrete buildings in the 22 February Christchurch (Lyttelton) earthquake. *Bulletin of the New Zealand Society for Earthquake Engineering* **44**(4): 239–278.

Makris, N. & Black, C.J. 2004. Dimensional analysis of rigid-plastic and elasto-plastic structures under pulse-type excitations. *Journal of Engineering Mechanics* **130**(9), 1006–1018.

Oddbjornsson, O., Alexander, N.A., Taylor, C.A. & Sigbjörnsson, R. 2012. Numerical and experimental exploration of the fundamental nonlinear dynamics of self-centring damage resistant structures under seismic excitation. in: 15th World Conference of Earthquake Engineering, Lisbon, # 4670, 10 pages.

Osinga, H.M. 2014. Computing failure boundaries by continuation of a two-point boundary value problem. in: Cunha, A., Caetano, E., Ribeiro, P., Muller, G. (Eds.) Proceedings of the 9th International Conference on Structural Dynamics, EURODYN 2014, Porto #MS10-263, pp. 1891–1897.

Priestley, M.J.N., Sritharan, S., Conley, J.R. & Pampanin, S. 1999. Preliminary results and conclusions from the PRESSS five-story precast concrete test building. *PCI Journal* **44**(6): 42–67.

Qin, X., Chen, Y. and Chouw, N. 2013. Effect of uplift and soil nonlinearity on plastic hinge development and induced vibrations in structures. *Advances in Structural Engineering* **16**(10): 135–147.

Ryan, H. 1994. Ricker, Ormsby, Klauder, Butterworth—A choice of wavelets. *CSEG Recorder* **19**(7): 8–9.

Takewaki, I., Moustafa, A. & Fujita, K. 2012. *Improving the Earthquake Resilience of Buildings: The Worst Case Approach*. London: Springer-Verlag.

Takewaki, I., & Kojima, K. 2017. Double, triple and multiple impulses for critical elastic-plastic earthquake response analysis to near-fault and long-duration ground motions. Chapter in this volume.

Modeling of soil-foundation-structure interaction for earthquake analysis of 3D BIM models

H. Werkle
University of Applied Sciences (HTWG), Konstanz, Germany

ABSTRACT: BIM as a new planning tool in construction facilitates the generation of three-dimensional finite element models of buildings. Hence, 3D FE models are available more and more also for earthquake analysis. The paper shows how dynamic Soil-Foundation-Structure-Interaction (SFSI) can be included in the analyses of 3D models in engineering practice with commercial FE software. Flexibility of the foundation slab is taken into account. A model of coupled distributed (Winkler) springs and dampers based on half-space solutions of impedance functions is suggested which consistently represents the rocking and vertical motions of the foundation. The results compare well with a more rigorous but numerically efficient SBFEM (Scaled Boundary Finite Element Method) analysis for a 3-storey building on a half-space.

1 INTRODUCTION

Building Information Modeling (BIM) as a new digital planning method of buildings gives an easy access to detailed 3D structural models of buildings for earthquake analysis. Modelling of SFSI, however, is not as obvious as for beam models of buildings typically used in an earthquake analysis so far. Spring and damper elements representing the dynamic impedance functions of a rigid foundation on soil imply some rigorous assumptions such as the missing deformability of the foundation slab. In the following, the modeling of dynamic SFSI in 3D building models is outlined including the foundation flexibility and the dynamic properties of the soil as, for example, radiation damping.

2 BUILDING INFORMATION MODELING

The main concept of Building Information Modeling is the consistent use of a unique comprehensive building database in the planning, construction and maintenance phase of a building. This implies the generation of a 3D digital model of the building with geometrical and nongeometrical data describing its properties. 4D modeling includes the description of the building process. In addition, a 5D model describes the costs and cash flows in each phase of the construction. Due to its great benefits and its ability to manage the data workflow between different construction industry stakeholders, BIM is often considered as a new paradigm in the planning process.

The BIM model is the basis for the generation of the structural model for the 3D finite element analysis (FEA) of the building which consists mainly of beam and shell elements. Many commercial software products for BIM modeling are based on IFC standard (Industry Foundation Classes) as e.g. AUTODESK Revit. They allow to generate a structural model, to export it to a commercial FEA software (supporting the interface) and to reimport the results. When it comes to an earthquake analysis this 3D model should be used for a consistent modeling workflow. It also represents the most accurate description of the structure. Simplified approaches like the application of equivalent static earthquake loads will not be considered here. Hence, in general SFSI with a flexible foundation plate and the dynamic stiffness and damping of soil has to be modeled for the 3D building finite element (FE) model. Especially concepts for modeling SFSI suited for practical purposes and applicable when using commercial FEA software, will be discussed.

3 SOIL-FOUNDATION-STRUCTURE INTERACTION IN EARTHQUAKE ANALYSIS

3.1 *General*

The main effects of SFSI are generally considered to be an elongation of the fundamental period of the building and an increase in damping. Mostly but not always this has a beneficial effect on the seismic performance of structures. SFSI is influenced by linear and nonlinear soil properties, the layering of the soil and varying soil profiles under the foundation, the geometry and flexibility of the foundation, radiation damping in soil and possibly the mutual interaction of the foundations of neighboring buildings. For large buildings, kinematic effects associated with scattering of incoming waves by the foundation may be of importance.

Although extensive research has been conducted over many decades (Kausel 2010, Roesset 2013, Lou et al. 2013) appropriate methods to take SFSI into account are often not implemented in engineering practice and commercial software for earthquake analysis of buildings. Only buildings with particularly high safety requirements like NPP structures or LNG tanks are usually analyzed with sophisticated and extensive methods including SFSI.

3.2 *Models and methods of analysis*

The analysis of SFSI is generally based on Green's functions describing the dynamic displacements of the soil caused by a point force acting at the soil surface. Green's functions have been given for a half-space as well as for layered soils mostly in the frequency domain (Kausel 1981, Waas et al. 1985). They may be used to construct a stiffness matrix of the soil related to the nodal points connecting the soil with the foundation slab of the building. This procedure for the analysis of layered soils in frequency domain has been implemented software specialized in SFSI analysis as e.g. SASSI (based on Lysmer et al. 1981). As large finite element models of the building including soil stiffness are computed in frequency domain, the method is cumbersome, intricate to use and requires large computing time, but it is very powerful.

Green's functions in frequency domain can be used to compute the dynamic impedance functions of rigid foundations, i.e. their frequency dependent springs and dampers. Solutions are available for circular and rectangular rigid foundations on a homogeneous half-space and on layered soils (Gazetas 1983, Sieffert et al. 1995). The Thin-Layer-Method according to Kausel (1981) and Waas et al. (1985) as well as the "direct stiffness method" by integration over wavenumbers as implemented in the EDT software tool by Schevels et al. (2010) are well suited to determine frequency dependent impedance functions for arbitrarily layered soils. In order to simulate the dynamic behavior of rigid foundations on a half-space, simple mass-spring-damper models have been developed (Wolf 1994).

Springs and dampers are widely used in the engineering community and there is a lot of experience in this field. For 3D FE models with a rigid foundation plate, an analysis in frequency domain with frequency dependent springs and dampers is another option. To model soil for non-rigid foundation slabs in 3D finite element models of buildings they may be used as distributed springs and dampers (DSD) similar to Winkler springs in a static analysis (NIST 2012). An alternative method of this approach will be discussed later.

Another concept for including SFSI in dynamic finite element analyses is by using numerical methods. Soil can be represented by a 3D finite element model with dampers at the boundaries in order to avoid artificial wave reflections as given first by Lysmer & Kuhlemeyer (1969). In frequency domain 3D transmitting, i.e. non-reflecting, boundaries have been given for a layered soil allowing for wave propagation to infinity (e.g. Werkle 1986, Werkle 1987). Other options are the Boundary Element Method (BEM) and the Scaled Boundary Finite

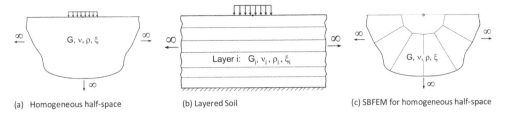

| (a) Homogeneous half-space | (b) Layered Soil | (c) SBFEM for homogeneous half-space |

Figure 1. Soil models.

Element Method (SBFEM) for an elastic half-space in frequency or in time domain (Wolf 2003). Some models are shown in Figure 1.

The focus in the structural analysis of 3D BIM models is on a detailed modeling of a building and on its structural behavior. Hence, extensive 3D FE models of the soil are not appropriate from an engineering point of view. This excludes large soil models with solid elements even when transmitting boundaries are used to limit the model size.

In this study the SBFEM in time domain as described by Radmanovic and Katz (2010) is used. As an alternative method, uncoupled or coupled distributed springs and dampers (DSD/EST) are discussed. The paper focuses on SSI effects in homogeneous soil, produced by the inertia of the superstructure. Kinematic effects are not addressed. All FE computations including SBFEM are performed using the finite element Software SOFiSTiK.

3.3 *Half-space models*

A classical soil model to analyze SFSI is the elastic or viscoelastic half-space. It should be noted however that even in a soil of homogeneous material, stiffness increases with depth due to the increase of the confining pressure. Therefore foundation impedances derived assuming homogeneous half-space conditions of soil often over-predict radiation damping compared to actual soil profiles. For soil with linearly increasing shear modulus 'equivalent depths' of an equivalent elastic half-space have been given by Werkle (1988). They depend on the motion of the foundation but in the case of dynamic impedances also on the wavelength or the frequency of excitation.

3.4 *DSD and EST models*

Buildings with a foundation slab typically stiffened by walls in the basement floor can be assumed to behave similarly as a rigid foundation slab, i.e. the deviations of the displacements of the foundation from rigid body motions can be neglected in the global soil stiffness and radiation damping. The impedance functions for a rigid foundation with radius r on an elastic half-space are written in frequency domain as

$$\tilde{K}_{(j)}(\omega) = K_{(j)}(\omega) + i \cdot \omega \cdot C_{(j)}(\omega) = K_{stat,(j)} \cdot \left(k_{(j)}(\omega) + i \cdot a_0 \cdot c_{(j)}(\omega) \right) \tag{1}$$

with the circular frequency of vibration $\omega = 2 \cdot \pi \cdot f$ and $a_0 = \omega \cdot r / v_s$. Here $v_s = \sqrt{G/\rho}$ (G = shear modulus, ρ = density) is the shear wave velocity in the half-space. The index (j) stands for 'h', 'r', 'v' i.e. for the horizontal, rocking and vertical vibration mode, respectively. The static spring constants are

$$K_{stat,h} = \frac{8 \cdot G \cdot r}{2 - v} \quad K_{stat,r} = \frac{8 \cdot G \cdot r^3}{3 \left(1 - v \right)} \quad K_{stat,v} = \frac{4 \cdot G \cdot r}{1 - v} \tag{2}$$

with the Poisson ratio v of the soil. The frequency dependent coefficients $k_{(j)}(\omega)$ and $c_{(j)}(\omega)$ are given e.g. by Gazetas (1983). Equation (1) can be understood as frequency dependent springs and dampers

$$K_{(j)}(\omega) = K_{stat,(j)} \cdot k_{(j)}(\omega), C_{(j)}(\omega) = K_{stat,(j)} \cdot c_{(j)}(\omega) \cdot \frac{a_0}{\omega}. \tag{3}$$

The springs and dampers of non-circular foundations can be approximated by a circular foundation having the same area A_F or the second moments of area $I_{F,\xi}$, $I_{F,\eta}$ for horizontal and rocking motion, respectively, as an equivalent circular foundation. A_F and $I_{F,\xi}$, $I_{F,\eta}$ relate to the contact area of the foundation and the soil and ξ, η are its principal axes (Fig. 2). Hence the equivalent radii $r_h = \sqrt{A_F/\pi}, r_v = \sqrt{A_F/\pi}, r_{r,\xi} = \sqrt[4]{4 \cdot I_{F,\xi}/\pi}, r_{r,\eta} = \sqrt[4]{4 \cdot I_{F,\eta}/\pi}$ are obtained for horizontal, vertical and rocking motion about the ξ- and η-axis, respectively.

For a three-dimensional model of a building, these global values are to be transformed to the nodal points of the finite element model of the foundation slab. Assuming a linear stress distribution of the soil stresses acting on the foundation, the transformation can be done by defining a distributed spring or modulus of subgrade reaction acc. to Winkler's model for horizontal and rocking motion

$$K_{SG,h}(\omega) = \frac{K_h(\omega)}{A_F}, \qquad K_{SG,r}(\omega) = \frac{K_r(\omega)}{I_F}. \tag{4}$$

The corresponding distributed damping is obtained as

$$C_{SG,h}(\omega) = \frac{C_h(\omega)}{A_F}, \qquad C_{SG,r}(\omega) = \frac{C_r(\omega)}{I_F}. \tag{5}$$

These distributed stiffness and damping (DSD) values are transformed to nodal point springs and dampers by standard finite element techniques (Fig. 5b).

In cases where vertical motion, rocking about two axes as well as horizontal motion in two directions and torsional motion are considered simultaneously, the soil reaction cannot be described by a unique subgrade modulus. For vertical motion e.g. the subgrade modulus is $K_{SG,v}(\omega) = K_v(\omega)/A_F$ whereas for rocking about the ξ-axis it is obtained to be $K_{SG,r,\xi}(\omega) = K_{r,\xi}(\omega)/I_{F,\xi}$ and about the η-axis $K_{SG,r,\eta}(\omega) = K_{r,\eta}(\omega)/I_{F,\eta}$.

In order to overcome this discrepancy, the equivalent stress transformation (EST) as given by Werkle (2008) can be applied, leading to a fully coupled stiffness matrix related to the nodal points of the foundation slab instead of a diagonal matrix as obtained for the DSD model. The method has also been applied successfully to the coupling of columns and plates in flat slabs (Werkle 2002).

The soil springs for vertical and rocking motion are given by

$$\underline{F}_{Sp} = \underline{K}_{Sp} \cdot \underline{w}_{Sp} \text{ with } \underline{F}_{Sp} = \begin{bmatrix} F_z \\ M_\eta \\ M_\xi \end{bmatrix}, \underline{w}_{Sp} = \begin{bmatrix} w_z \\ \varphi_\eta \\ \varphi_\xi \end{bmatrix}, \underline{K}_{Sp} = \begin{bmatrix} k_z & 0 & 0 \\ 0 & k_{r,\eta} & 0 \\ 0 & 0 & k_{r,\xi} \end{bmatrix}. \tag{6}$$

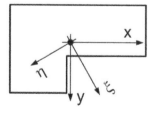

Figure 2. Foundation slab.

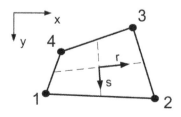

Figure 3. Plate element.

102

In the EST a linear distribution of the contact stresses under the foundation is assumed as

$$p_z(\xi, \eta) = \frac{F_z}{A_z} - \frac{M_\eta}{I_\eta} \cdot \xi + \frac{M_\xi}{I_\xi} \cdot \eta. \tag{7}$$

The stresses at the n nodal points of the foundation slab with coordinates ξ_j, η_j are

$$\underline{p}_{Pl} = \underline{\Xi} \cdot \underline{I} \cdot \underline{F}_{Sp} \text{ with } \underline{p}_{Pl} = \begin{bmatrix} p_1 \\ p_2 \\ p_3 \\ . \\ . \\ p_n \end{bmatrix}, \underline{\Xi} = \begin{bmatrix} 1 & \xi_1 & \eta_1 \\ 1 & \xi_2 & \eta_2 \\ 1 & \xi_3 & \eta_3 \\ . & . & . \\ . & . & . \\ 1 & \xi_n & \eta_n \end{bmatrix}, \underline{I} = \begin{bmatrix} 1/A_z & 0 & 0 \\ 0 & -1/I_\eta & 0 \\ 0 & 0 & 1/I_\xi \end{bmatrix} \tag{8}$$

These soil pressures $\underline{p}_{Pl}$ are applied as linearly "distributed loads" to the finite elements of the slab and transformed to equivalent nodal forces applying the principle of virtual displacements. For a 4-node plate element (Fig. 3) one obtains the nodal forces $\underline{F}^{(el)}$ perpendicular to its plane in element coordinates r, s ([–1 1]) as

$$\underline{F}^{(el)} = \int \underline{N} \cdot p \cdot dA, \underline{F}^{(el)} = \begin{bmatrix} F_{z1} \\ F_{z2} \\ F_{z3} \\ F_{z4} \end{bmatrix}, \underline{N}(r,s) = \frac{1}{4} \cdot \begin{bmatrix} (1-r)\cdot(1+s) \\ (1+r)\cdot(1+s) \\ (1+r)\cdot(1-s) \\ (1-r)\cdot(1-s) \end{bmatrix}, dA = Det(\underline{J}) \cdot dr \cdot ds. \tag{9}$$

with $Det(\underline{J})$ as determinant of the Jacobi operator. The stresses p, applied at the plate elements can be interpolated from their nodal values p_i as

$$p(r,s) = \underline{N}(r,s)^T \cdot \underline{p}^{(el)} \text{ with } \underline{p}^{(el)T} = [p_1 \quad p_2 \quad p_3 \quad p_4] \tag{10}$$

For this bilinear distribution of p one obtains

$$\underline{F}^{(el)} = \underline{A}^{(el)} \cdot \underline{p}^{(el)} \text{ with } \underline{A}^{(el)} = \iint \underline{N} \cdot \underline{N}^T \cdot Det(\underline{J}) \, dr \, ds \tag{11}$$

The integral can be computed numerically by Gauss integration formula. However, for a rectangular element with the side lengths a, b in x- and y-direction respectively, one obtains

$$\underline{A}^{(el)} = \frac{a \cdot b}{36} \begin{bmatrix} 4 & 2 & 1 & 2 \\ 2 & 4 & 2 & 1 \\ 1 & 2 & 4 & 2 \\ 2 & 1 & 2 & 4 \end{bmatrix}. \tag{12}$$

The element forces $\underline{F}^{(el)}$ and the applied stresses $\underline{p}^{(el)}$ of an individual element are related to the nodal points of the foundation slab by the topology matrix $\underline{Z}^{(el)}$ (containing ones or zeros):

$$\underline{F}_{Pl}^{(el)} = \underline{Z}^{(el)T} \cdot \underline{F}^{(el)} \text{ and } \underline{p}^{(el)} = \underline{Z}^{(el)} \cdot \underline{p}_{Pl} \tag{13}$$

Summing up the contributions for all elements one obtains the forces $\underline{F}_{Pl}$ and the stresses $\underline{p}_{Pl}$ at nodal points of the plate with eq. (11) as

$$\underline{F}_{Pl} = \underline{A} \cdot \underline{p}_{Pl} \text{ with } \underline{A} = \sum_{(el)} \underline{Z}^{(el)T} \cdot \underline{A}^{(el)} \cdot \underline{Z}^{(el)}. \tag{14}$$

With (9) the transformation of the spring forces to nodal forces of the foundation slab is obtained to be

$$\underline{F}_{Pl} = \underline{T}^T \cdot \underline{F}_{Sp} \text{ with } \underline{T}^T = \underline{A} \cdot \underline{\Xi} \cdot \underline{I}. \tag{15}$$

For the transformation of the displacements and the rotations of the soil springs a corresponding transformation is valid, i.e.

$$\underline{w}_{Sp} = \underline{T} \cdot \underline{w}_{Pl}, \underline{w}_{Pl} = \begin{bmatrix} w_1 & w_2 & \text{........} & w_n \end{bmatrix}^T, \tag{16}$$

where $\underline{w}_{Pl}$ denotes the displacements of the nodal points of the plate.

The stiffness of the soil springs according to Eq. (6) can now be transformed onto the nodal points of the plate. Using Eq. (15) and Eq. (16) one obtains

$$\underline{F}_{Pl} = \underline{K}_{Pl} \cdot \underline{w}_{Pl} \text{ with } \underline{K}_{Pl} = \underline{T}^T \cdot \underline{K}_{Sp} \cdot \underline{T}. \tag{17}$$

Here $\underline{K}_{Pl}$ is the stiffness matrix of the soil related to the nodal points of the foundation slab.

With (6) and (16) the spring forces are obtained to be

$$\underline{F}_{Sp} = \underline{K}_{Sp} \cdot \underline{w}_{Sp} = \underline{K}_{Sp} \cdot \underline{T} \cdot \underline{w}_{Pl}. \tag{18}$$

As a numerical example a foundation slab is discretised in 6×4 elements (Fig. 4). Hence the number of nodal points is $5 \cdot 7 = 35$. The transformation matrix $\underline{T}$ related to the vertical degrees of freedom of the foundation slab acc. to the numbering scheme in Figure 4 is obtained as

$$\underline{T} = \frac{1}{288} \cdot \begin{bmatrix} 3 & 6 & 6 & 6 & 6 & 6 & 3 & 6 & 12 & 12 & 12 & 12 & 12 & 6 & 6 & 12 & 12 \\ \frac{16}{c} & \frac{24}{c} & \frac{12}{c} & 0 & -\frac{12}{c} & -\frac{24}{c} & \frac{16}{c} & \frac{32}{c} & \frac{48}{c} & \frac{24}{c} & 0 & -\frac{24}{c} & -\frac{48}{c} & \frac{32}{c} & \frac{32}{c} & \frac{48}{c} & \frac{24}{c} \\ \frac{15}{d} & \frac{30}{d} & \frac{30}{d} & \frac{30}{d} & \frac{30}{d} & \frac{30}{d} & \frac{15}{d} & \frac{18}{d} & \frac{36}{d} & \frac{36}{d} & \frac{36}{d} & \frac{36}{d} & \frac{36}{d} & \frac{18}{d} & 0 & 0 & 0 \end{bmatrix}$$

$$\begin{matrix} 12 & 12 & 12 & 6 & 6 & 12 & 12 & 12 & 12 & 12 & 6 & 3 & 6 & 6 & 6 & 6 & 6 & 3 \\ 0 & -\frac{24}{c} & -\frac{48}{c} & -\frac{32}{c} & \frac{32}{c} & \frac{48}{c} & \frac{24}{c} & 0 & -\frac{24}{c} & -\frac{48}{c} & -\frac{32}{c} & \frac{16}{c} & \frac{24}{c} & \frac{12}{c} & 0 & -\frac{12}{c} & -\frac{24}{c} & -\frac{16}{c} \\ 0 & 0 & 0 & 0 & -\frac{18}{d} & -\frac{36}{d} & -\frac{36}{d} & -\frac{36}{d} & -\frac{36}{d} & -\frac{36}{d} & -\frac{18}{d} & \frac{15}{d} & \frac{30}{d} & \frac{30}{d} & \frac{30}{d} & \frac{30}{d} & \frac{30}{d} & \frac{15}{d} \end{matrix} \end{bmatrix}$$

$$\tag{19}$$

It can be seen that the displacements of the spring according to (19) represent a weighted average of the nodal displacements of the slab. For rigid body motions (6) is fulfilled exactly.

Transformation matrices for the coupled motion of the horizontal motions in ξ- and η-direction and torsion can be derived by the same procedure. It's not necessary to use (frequency-dependent) spring models acc. to (6). Instead, models to represent SFSI in time domain by a mass-spring-damper model as given for example by Wolf (1994) may be coupled to the foundation slap by the transformation described above.

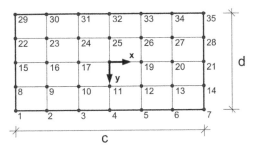

Figure 4.   FE discretization of a rectangular foundation plate with 6×4 elements.

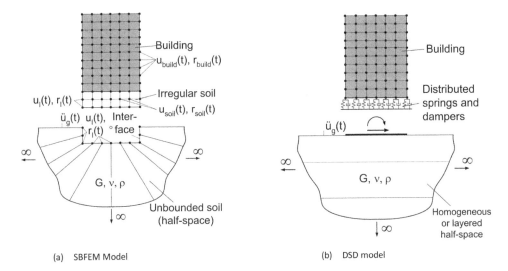

(a)   SBFEM Model

(b)   DSD model

Figure 5.   Models for soil-foundation-stucture interaction.

3.5   *SBFEM model*

For time history analyses the modelling of soil by the 3D SBFEM in time domain for an elastic half-space is straightforward (Fig. 5a). The irregular soil region under the foundation is discretized in finite solid elements with arbitrary material parameters. In the SBFEM the forces at the interface of the unbounded and the irregular soil region are given in time domain by

$$r_I(t) = \int_0^t M_I^\infty \cdot \ddot{u}_I(t-\tau)\, d\tau \tag{20}$$

where M_I^∞ is the acceleration unit-impulse matrix (Radmanovic, 2009). The equations of motion of the finite element model are (Fig. 5a)

$$\begin{bmatrix} M_{SS} & M_{SI} \\ M_{IS} & M_{II} \end{bmatrix} \cdot \begin{bmatrix} \ddot{u}_S(t) \\ \ddot{u}_I(t) \end{bmatrix} + \begin{bmatrix} C_{SS} & C_{SI} \\ C_{IS} & C_{II} \end{bmatrix} \cdot \begin{bmatrix} \dot{u}_S(t) \\ \dot{u}_I(t) \end{bmatrix} + \begin{bmatrix} K_{SS} & K_{SI} \\ K_{IS} & K_{II} \end{bmatrix} \cdot \begin{bmatrix} u_S(t) \\ u_I(t) \end{bmatrix} = \begin{bmatrix} p_S(t) \\ 0 \end{bmatrix} + \begin{bmatrix} 0 \\ r_I(t) \end{bmatrix} \tag{21}$$

The mass, damping and the stiffness matrices are subdivided in the degrees of freedom relating to the structure and to the interface. The structure consists of the irregular region of the soil and of the building both discretized in finite elements, hence

$$u_S(t) = \begin{bmatrix} u_{Build}(t) \\ u_{Soil}(t) \end{bmatrix}, \; p_S(t) = \begin{bmatrix} p_{Build}(t) \\ 0 \end{bmatrix}. \tag{22}$$

The earthquake acceleration time history is defined at the soil surface. Therefore

$$p_{Build}(t) = - M_{Build} \cdot I_x \cdot \ddot{u}_g(t) \tag{23}$$

where M_{Build} is the mass matrix of the building and I_x the influence vector (elements are "1" in the direction of the earthquake acceleration and "0" in all other degrees of freedom).

It should be noted that the SBFEM in time domain has been developed for an elastic half-space without internal material damping, i.e. only radiation damping is considered in the analysis.

4 EXAMPLE

4.1 *Structural model*

A three story building over a basement floor has been investigated (Fig. 6). The foundation slab has a size of 18×21 m, the total height of the building is 13 m. A floor plan of the first floor is shown in Figure 6a. The walls have a thickness of 0.30 m and columns a size of 0.3×0.3 m. The slab thicknesses are 0.20 m and the foundation slab thickness is 0.40 m. The building consists of concrete with a Young's Modulus of E = 28300 MN/m² and a material damping of 5%. In this study, only earthquake action in y-direction is considered where the building is symmetric.

The soil consists of soft clay with a shear modulus of 30 MN/m² and a Poisson ratio of 0.33 i.e. $v_S = 129 \, m/s$. Its density is 1.8 to/m³. Material damping ξ_{soil} in soil is neglected.

Earthquake action is defined at the interface between the soil and the structure, i.e. at the soil surface. An elastic response spectrum acc. to Eurocode 8 (DIN EN 1998-1/German NA, 2011) for a ground type C-S, a ground peak acceleration $a_{gR} = 0.8 \, m/s^2$ and importance factor $\gamma_I = 1.0$ is assumed. For the time history analyses the mean value of four spectrum compatible time histories are evaluated.

4.2 *Computational investigations*

The influence of soil-structure interaction and of the flexibility of the foundation slab on the global response of the structure has been studied with a model with a fixed base (without

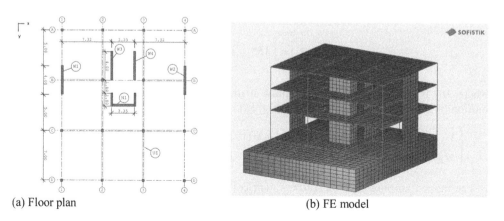

(a) Floor plan   (b) FE model

Figure 6. Three-dimensional finite element model of a multi-storey building (Volarevic, 2013).

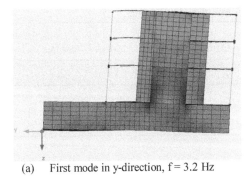
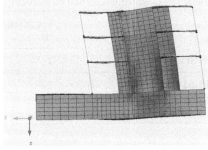

(a)   First mode in y-direction, f = 3.2 Hz   (b)   Second mode in y-direction, f = 8.3 Hz

Figure 7.   Mode shapes of the DSD model with flexible foundation slab.

SFSI), and two models including SFSI, one with rigid and another with a flexible foundation slab. For the two models with SFSI, time history analyses (THA) have been performed with DSD models by modal analysis as well as with the SBFEM models by direct integration. In addition a response spectrum analysis (RSA) has been done for the DSD models.

The modelling of damping in SFSI is a major issue, still not completely understood. In this study in response spectrum analysis, the different damping behavior of soil and building is taken into account approximately as modal damping. The modal damping coefficient ξ_i in the i-th mode has been obtained as weighted average of the damping coefficients of the building and of the soil springs or dampers (Roesset et al 1973, Tsai 1974). The weighting factors are the potential energy in the building and in soil springs for horizontal and rocking motion, respectively, related to the total potential energy in the i-th mode. The damping coefficients ξ_h for the horizontal and ξ_r rocking motion of a rigid foundation are

$$\xi_h = \frac{a_{0,h}}{2} \cdot \frac{c_h(\omega_1)}{k_h(\omega_1)} + \xi_{soil}, \xi_r = \frac{a_{0,r}}{2} \cdot \frac{c_r(\omega_1)}{k_r(\omega_1)} + \xi_{soil}. \tag{24}$$

with $a_{0,h} = \omega_1 \cdot r_h / v_S$ and $a_{0,r} = \omega_1 \cdot r_r / v_S$ i.e. the impedance functions are evaluated at the first eigenfrequency of the soil-structure system. The damping coefficients given in Table 1 show that the damping for horizontal motion is very high and will almost prevent this type of motion. Modal damping has been limited to 28% due to approximate character of the method.

For direct integration in the time history analyses structural damping is modelled as Rayleigh damping whereas the Rayleigh coefficients are adapted to the material damping at the first two frequencies as given in Figure 7.

4.3   Structural response

The eigenfrequencies are given in Table 1. The first two (global) modes are considered since their modal masses represent approximately the total mass of the building. Mode shapes of the model with the flexible foundation slab are given in Figure 7.

The maximum accelerations, plotted at the shear centre of the floors over the building height, are shown in Figure 8. They show the significant influence of soil-structure interaction due to the soft soil conditions when compared to a model with a fixed base. For a flexible foundation slab larger accelerations are found than for a rigid plate. The DSD models give somewhat larger accelerations as the more rigorous SBFEM models. The results of the EST models are similar to the DSD models in this case since there is no simultaneous vertical earthquake excitation.

Local stress resultants as normal forces and shear forces in the walls or bending moments in the foundation slabs are obtained with all models (Werkle & Volarevic 2014).

Table 1. Eigenfequencies and modal damping coefficients of the DSD model.

SFSI model	Eigenfrequencies [Hz] 1.	2.	Damping in soil ξ_h	ξ_r	Modal damping 1.	2.
Fixed base (without SFSI)	5.8	23.4	–	–	–	–
Rigid foundation slab	3.7	8.6	58%	33%	27%	28% (30%)
Flexible foundation slab	3.2	8.3	50%	25%	17%	27%

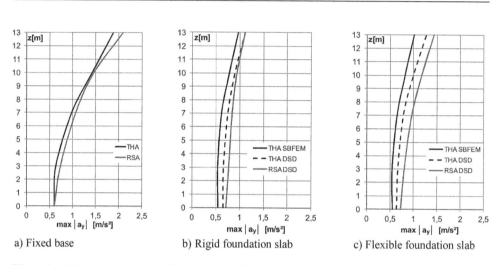

a) Fixed base   b) Rigid foundation slab   c) Flexible foundation slab

Figure 8.   Mode shapes of the DSD model with flexible foundation slab.

5   CONCLUSIONS

Three-dimensional FE models of buildings coming up more and more with the new BIM concepts in structural planning should take SFSI into account. For this the implementation of additional tools in commercial software such as DSD/EST models are suggested. Research is required to solve SFSI problems with flexible foundation slabs efficiently in time domain such as SBFEM methods for realistic soil models as layered soils or soils with a stiffness increasing with depth.

REFERENCES

Gazetas, G. 1983. Analysis of machine foundation vibrations: state of the art. *Soil Dynamics and Earthquake Engineering* 2 (1). 2–42.

Kausel, E. 1981. *An explicit solution for the Green's function for dynamic loads in layered media.* Research Report R81-13, Massachusetts Institute of Technology, Boston.

Kausel, E. 2010. Early history of soil-structure interaction. *Soil Dynamics and Earthquake Engineering* 30: 822–832.

Lou M., Wang, H., Chen, X. & Zhai, Y. 2011. Structure-soil-structure interaction: Literature review. *Soil Dynamics and Earthquake Engineering* 31: 1724–1731.

Lysmer, J. & Kuhlemeyer, R.L. 1969. Finite dynamic model for infinite media. *Journal of the Engineering Mechanics Division* 95(4), ASCE. 859–877.

Lysmer, J., Tabatabaie-Raissi, M., Tajirian, F., Vahdani, S. & Ostadan, F. 1981. *SASSI-A System for Analysis of Soil-Structure Interaction.* Report no. UCB/GT/81-02, Department of Civil Engineering, University of California Berkeley.

NIST. 2012. *Soil-structure interaction for building structures.* NIST GCR 12-917-21, US Department of Commerce.

Radmanovic, B. & Katz, C. 2010. High Performance SBFEM. *Proc. of the World Congress of Computational Mechanics (WCCM)*, Sydney.

Roesset, J.M. Whitman, R., Dobry, R. 1973. Modal analysis for structures with foundation interaction. *Journal of the Structrual Division* 99(3), ASCE. 399–416.

Roesset, J. 2013. Soil structure interaction—the early stages. *Journal of Applied Science and Engineering* 16 (1): 1–8.

Schewels M., Francois, S. & Degrande, G. 2010. *EDT Elastodynamic Toolbox for Matlab*, Users guide, Report BWM-2010-11, Katholieke Universiteit Leuven.

Sieffert, J.G. & Cevaer, J. 1995. *Handbook of impedance functions.* Rennes: Editions Ouest-France.

Sofistik, *ASE Manual*, Vers. 2014-9, Sofistik AG, Oberschleißheim, Germany, 2014.

Tsai, N.C. 1974. Modal damping for soil-structure interaction. *Journal of the Engineering Mechanics Division* 100(2), ASCE: 323–341.

Volarevic, J. 2013. *Boden-Bauwerk-Wechselwirkung bei der dynamischen Finite-Element-Berechnung von Gesamtmodellen.* Master Thesis. Hochschule Technik Wirtschaft und Gestaltung Konstanz (HTWG). Konstanz (in German).

Waas, G., Riggs, R.H. & Werkle, H. 1985. Displacement solutions for dynamic loads in a transversely-isotropic stratified medium. *Earthquake Engineering and Structural Dynamics* 13, 173–193.

Werkle, H. 1986. Dynamic finite element analysis of three-dimensional soil models with a transmitting element. *Earthquake Engineering and Structural Dynamics* 14. 41–60.

Werkle, H. 1987. A transmitting boundary for the dynamic finite element analysis of cross anisotropic soils. *Earthquake Engineering and Structural Dynamics* 15. 831–838.

Werkle, H. 1988. Steifigkeit und Dämpfung von Fundamenten auf inhomogenem Baugrund. In: Steinwachs, M.: *Ausbreitung von Erschütterungen im Boden und Bauwerk.* 3. Jahrestagung der DGEB, Clausthal: Trans Tech Publications (in German).

Werkle, H. 2002. Analysis of flat slabs by the finite element method. *2nd International Symposium on Advances in Structural Engineering and Mechanics* ASEM'02, Busan.

Werkle, H. 2008. *Finite Elemente in der Baustatik.* 3rd ed., Wiesbaden: Vieweg (in German, English edition planned to be published by Springer in 2018).

Werkle, H. & Volarevic, J. 2014. Modeling of dynamic soil-structure interaction in the three-dimensional finite element analysis of buildings. *Proc. 15th European Conference on Earthquake Engineering, Istanbul.*

Wolf, J.P. 1985. *Dynamic soil structure interaction.* Englewood Cliffs N.J.: Prentice Hall.

Wolf, J.P. 1994. *Foundation vibration analysis using simple physical models.* Englewood Cliffs N.J.: Prentice Hall.

Wolf J.P. 2003. *The Scaled Boundary Finite Element Method.* Chichester: John Wiley and Sons.

Seismic Performance of Soil-Foundation-Structure Systems – Chouw, Orense & Larkin (Eds)
© 2017 Taylor & Francis Group, London, ISBN 978-1-138-06251-1

Centrifuge modelling of the seismic response of multi-storey buildings on raft foundations to the Christchurch Earthquake

L.B. Storie
Tonkin + Taylor Ltd., Auckland, New Zealand
University of Auckland, Auckland, New Zealand

M.J. Pender
University of Auckland, Auckland, New Zealand

J.A. Knappett
University of Dundee, Dundee, UK

ABSTRACT: There were a number of multi-storey buildings on shallow raft foundations in the Central Business District (CBD) of Christchurch that performed well in the 22 February 2011 Christchurch Earthquake. Structural assessments following the earthquake have concluded that some buildings performed significantly better than would have been expected given the intensity of the recorded ground motions in and around the central city. Nonlinear Soil-Foundation-Structure Interaction (SFSI) provides a possible explanation for the good performance of these buildings. Centrifuge experiments were undertaken at the University of Dundee, U.K., to examine the influence of SFSI in the seismic response of multi-storey buildings on raft foundations using a range of equivalent Single Degree of Freedom (SDOF) building models resting on a layer of dense, dry sand. The models were subjected to representative records from the Christchurch Earthquake and it was found that significant energy was dissipated between the soil, foundation and structure. The large raft in conjunction with dense sand meant significant energy could be dissipated through SFSI without the detrimental effects of significant permanent soil deformation.

1 INTRODUCTION

The interaction between the soil, foundation and structure during an earthquake has the potential to significantly influence the seismic performance of multi-story buildings on shallow foundations. Typically, this interaction is neglected in current analysis and design practice, and the foundation is assumed to be fixed to the ground. However, during large earthquake shaking there is often not enough vertical load to preclude uplift of a shallow foundation and this can lead to plastic deformation of the underlying soil (Martin and Lam, 2000, Kelly, 2009). Experimental and analytical studies have found a reduction in the forces transmitted to structures whose foundations are able to uplift from the supporting soil but highlight the possibility of detrimental settlement due to permanent soil deformation (Harden et al., 2005, Pecker and Chatzigogos, 2010, Gajan et al., 2010, Anastasopoulos, 2010, Loli et al., 2014). The term Soil-Foundation-Structure Interaction (SFSI) has been coined to describe these nonlinear geometrical and nonlinear soil deformation effects at the soil-foundation interface, differentiating the process from classical linear elastic Soil-Structure Interaction (SSI) (Orense et al., 2010).

SFSI incorporates nonlinear geometrical effects and possible nonlinear soil deformation effects into earthquake analysis of structure-foundation systems. For shallow foundations, this involves uplift of the foundation and possible plastic deformation of the underlying soil during large earthquake shaking. These effects provide a possible explanation for the good performance of a number of buildings on shallow foundations in the city of Christchurch, New Zealand during the Christchurch Earthquake of February 22, 2011. Structural assessments following the earthquake have concluded that some buildings have performed significantly better than would have been expected given the intensity of the recorded ground motions in and around the Christchurch central business district (CBD). These buildings generally had large raft foundations resting on a thick layer of competent gravel so had a high static bearing capacity factor of safety (FoS) and were not affected by liquefaction. It is suggested that foundation uplift and possible plastic soil deformation have reduced the forces transmitted to these structures. Centrifuge experiments have been undertaken to assess the potential influence of SFSI in the seismic response of these buildings on shallow foundations.

Centrifuge experiments were conducted at the University of Dundee, U.K., to investigate the rocking response of a range of multi-storey buildings on raft foundations resting on a layer of dense sand. Centrifuge modelling is important when investigating soil-foundation systems because confining stresses play an important role in soil constitutive behavior (Taylor, 1995). Equivalent elastic single degree of freedom (SDOF) building models were used in the experiments since the focus was the nonlinear mechanisms at the soil-foundation interface and how those mechanisms influenced overall building response. Equivalent 3, 5, and 7 storey building models with identical sized square raft foundations were designed for the experiments and were placed on the surface of a prepared layer of dense, dry sand. The sand represented an idealisation of a non-liquefiable cohesionless deposit. The combination of large raft foundation and dense sand meant that the static bearing capacity FoS for the models was large, simulating the scenario observed in Christchurch. The models were then subjected to representative Christchurch Earthquake records to investigate the extent of potential SFSI in the seismic response of multi-storey buildings on shallow foundations in Christchurch.

2 MULTI-STOREY BUILDINGS ON SHALLOW FOUNDATIONS IN THE CHRISTCHURCH EARTHQUAKE

Following the 4 September 2010 moment magnitude (M_w) 7.1 Darfield Earthquake, a large series of aftershocks affected the city of Christchurch, New Zealand. In particular, the M_w 6.2 Christchurch Earthquake on 22 February 2011 significantly impacted that part of the country. That aftershock is the most costly earthquake to have affected New Zealand, with 181 fatalities and significant damage to lifelines, infrastructure and residential and commercial buildings (Cubrinovski et al., 2011). Despite the tragedy of this earthquake, a wealth of data on the response of the built environment has been gathered, allowing insights into the behavior of structures, their foundations and the soil they are founded on.

In the CBD of Christchurch there are a number of multi-story buildings on shallow foundations that have performed satisfactorily despite the strong levels of ground shaking (Storie, 2017). Seven particular buildings of interest have been identified and are presented in Table 1. These buildings are in areas where there was no significant manifestation of liquefaction at the ground surface in the Christchurch Earthquake. Shallow foundations for these buildings often involved a basement, typically one storey beneath the ground surface, and a thick raft foundation. Geotechnical subsurface investigations at these building locations are presented in Figure 1 and show the basements would be founded on a thick layer of dense sandy gravel between 1 and 4 meters below the ground surface. The combination of a large raft foundation on a layer of competent gravel meant that these buildings would have a high static bearing capacity FoS.

Structural assessments of these buildings using a traditional fixed base approach may suggest that some buildings have performed better than expected in the Christchurch Earthquake. One such building, the 11 story steel-framed HSBC building that sits on a thick shallow raft foundation, was surveyed following the earthquake and was found to have self-centered to within construction tolerances, being a maximum of 0.07° off vertical (Bolland, 2011). Comparison of the elastic design spectrum from NZS1170.5 (Standards New Zealand, 2004) with the earthquake spectrum from the nearest strong ground motion stations

Table 1. Buildings of interest that performed satisfactorily in the Christchurch Earthquake and are on shallow foundations.

Building code	Building name	Street address	Number of storeys
A	Forsyth Barr Building	764 Colombo Street	18
B	Victoria Square Carpark	102 Armagh Street	4
C	Novatel Christchurch	52 Cathedral Square	14
D	HSBC Building	62 Worcester Street	11
E	Ibis Hotel	107 Hereford Street	9
F	Scorpio Books	79–83 Hereford Street	7
G	Lichfield Carpark	29–33 Lichfield Street	5

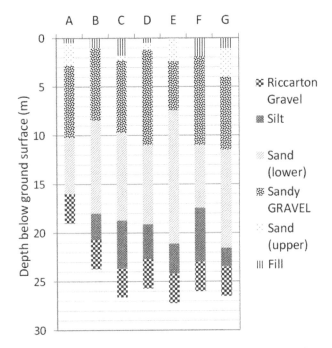

Figure 1. Subsurface ground conditions at buildings of interest (see Table 1) in the Christchurch CBD.

in the Christchurch CBD showed that the earthquake was around 2 times the ultimate limit state event that the building was designed for. Combining this with the calculated ultimate design drift and comparing with measured scuff marks from the movement of stairs within the building, it was assessed that the building response appeared to be more than twice as stiff as expected (Clifton, 2013). However, it is postulated that rather than the building being stiffer, SFSI effects meant that the forces transferred to the structure were reduced.

3 CENTRIFUGE MODELLING

Centrifuge experiments were undertaken at the University of Dundee, U.K. on soil-foundation-structure models designed to represent the scenario in Christchurch where multi-storey buildings on shallow raft foundations performed satisfactorily. Considering the space limitations in the centrifuge, generic frame structures were developed for 3, 5, and 7 storey buildings using assumptions regarding overall building dimensions and mass distribution. An example of the 7 storey frame structure developed for the centrifuge experiments is presented in Figure 2 and all the models were designed to have the same size square raft foundation

(8 meters square). Procedures outlined in Chapter 3 of Priestley et al. (2007) were followed to develop equivalent SDOF substitute structures for the generic buildings.

The calculated prototype scale lumped mass, effective height, and assumed fixed base natural period of the three building models are presented in Table 2 and these were used to develop scale models for the centrifuge experiments conducted at a centripetal acceleration of 50 g. Thus a scale factor of 50 was used when developing the scale models and when scaling the time histories (refer to Schofield, 1980). An example cross section and three dimensional drawing of the 7 storey structure-foundation model used in the centrifuge experiments (i.e. at model scale) are presented in Figure 3. The column dimensions in the SDOF models were chosen to achieve the fixed base period given the size of lumped structure mass at the top of the column. The mass of the foundation at prototype scale could also be determined using the dimensions of the model and then the total vertical load of the equivalent SDOF model was used with the properties of the soil to determine the static bearing strength factor of safety (FoS), also presented in Table 2. As can be seen, the static bearing capacity FoS values were quite large due to a large raft foundation on a competent soil, analogous to the scenario in Christchurch.

In each case the model structures were placed atop a layer of dry Congleton silica sand ($\rho_{dmin} = 1487$ kg/m³, $\rho_{dmax} = 1792$ kg/m³, $D_{60} = 0.14$ mm, $f_{crit} = 32$ degrees), which was prepared uniformly by air pluviation to have a relative density of approximately 83%. The deposit of sand was 200 mm deep, equivalent to 10 m at prototype scale given the centripetal acceleration of 50 g resulting in the scale factor of 50, and was prepared within an equivalent shear beam container that minimises dynamic boundary effects (Bertalot, 2013). Instrumentation consisted of 9 accelerometers (±70 g range) within the soil and 7 on each structure-foundation

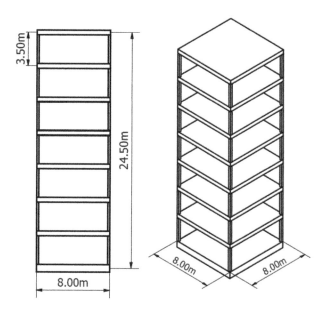

Figure 2. Generic 7 storey frame building developed for application in the centrifuge experiments.

Table 2. Prototype scale equivalent SDOF building characteristics for the three models tested in the centrifuge experiments.

Building model	Structure mass (T)	Effective height (m)	Fixed base period (s)	Foundation mass (T)	Total vertical load (kN)	Static bearing capacity FoS
3 storey	109.2	7.4	0.3	173.4	2920.4	92.6
5 storey	187.2	11.9	0.5	174.1	3969.1	68.1
7 storey	259.0	16.5	0.7	170.5	5010.9	54.0

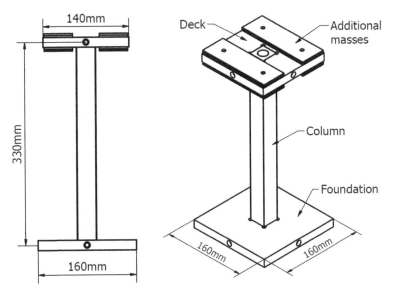

Figure 3. Equivalent SDOF structure-foundation model of the generic 7 storey building used in the centrifuge experiments (dimensions at model scale).

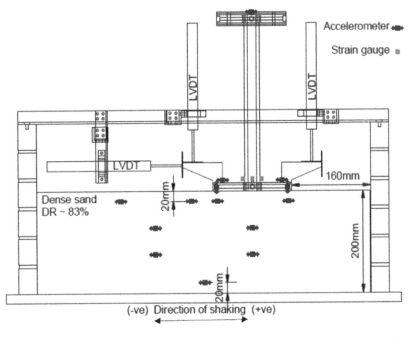

Figure 4. Layout of the centrifuge experiment for the 7 storey model on the layer of dense sand (dimensions at model scale).

model, along with two pairs of strain gauges at the base of the column of each of the equivalent SDOF models to measure bending moments input to the foundations. Linear Variable Differential Transformers (LVDTs) were connected to the foundation to measure vertical, horizontal and rotational displacement. The layout of the experiments is presented in Figure 4, with the 7 storey model shown for reference.

4 CHRISTCHURCH EARTHQUAKE EXPERIMENTAL RESULTS

The three equivalent SDOF building models on the layer of dense sand were spun up to 50 g in the centrifuge and then subjected to Christchurch Earthquake strong ground motion records, appropriately scaled (Schofield, 1980). Strong ground motion recording stations in the vicinity of the buildings of interest and with subsurface soil profiles similar to that present at the buildings of interest were selected. The results from the CBGS strong ground motion recording station are primarily presented in this paper and a time history of the record and corresponding response spectra are presented in Figure 5. Similar results were obtained from the two other strong ground motion stations used (Storie, 2017).

4.1 *Structure-foundation response*

Direct measurements from the centrifuge experiments provided insights into the structure-foundation response and the influence of nonlinear SFSI. Vertical displacements of the edges of the foundation measured by the LVDTs are presented in Figure 6 for the three building models subjected to the CBGS Christchurch Earthquake record. The average of these displacements represents the displacement of the centre of the assumed rigid foundation. Uplift was considered to be occurring when positive displacement of the edges was noticeably more positive than the average because soil, particularly dry sand, is considered to have no tensile capacity.

Uplift of the edges of the foundation was considered to have occurred for all of the building models. However, the extent of uplift was small relative to the size of the foundation and ranged between approximately 10 mm and 40 mm in the experiments. This corresponds to between 0.1% and 0.5% of the width of the foundation.

Residual settlement of the foundation could also be observed at the end of the experiments and indicated permanent soil deformation had occurred. The extent of permanent settlement, as with uplift, was small compared to the size of the foundation, with a maximum settlement of approximately 10 mm across the experiments. Differences in settlement of opposite edges of the foundation, particularly noticeable in the 5 storey model results in Figure 6, resulted in very small residual rotations, with a maximum value of 0.1 degrees residual rotation across the experiments.

Acceleration was recorded within the soil and at points on the structure as outlined by the accelerometer locations in Figure 4. Comparisons are made between the input acceleration in the soil just beneath the foundation and on the top of the structure for the three building models subjected to the CBGS Christchurch Earthquake record in Figure 7. For the 7 storey model, only approximately 25% of the peak magnitude of the input was transferred to the

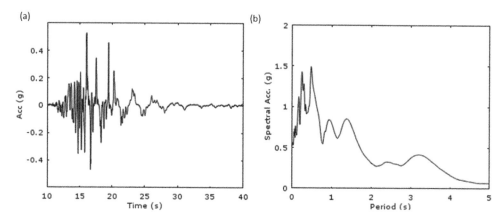

Figure 5.   Christchurch Earthquake (a) time history and (b) response spectra from the CBGS strong ground motion station.

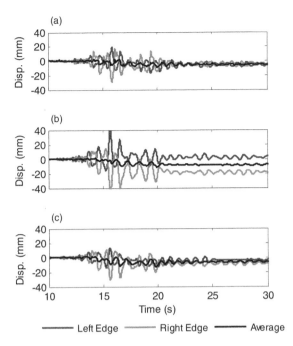

Figure 6. Foundation vertical displacement for the (a) 3 Storey, (b) 5 Storey, and (c) 7 Storey models subjected to the CBGS Christchurch Earthquake record.

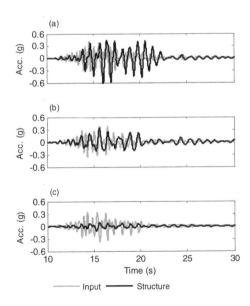

Figure 7. Acceleration time histories of the soil and structure for experiments on the (a) 3 storey, (b) 5 storey, and (c) 7 storey models subjected to the CBGS Christchurch Earthquake record.

structure. For the 5 storey model there was a slight attenuation and for the 3 storey model the peak response was amplified by approximately 55%. Nonlinear interaction at the soil-foundation interface (small extents of uplift and permanent soil deformation as shown in Figure 6) appeared to have a beneficial effect on the taller building models subjected to the

117

Christchurch Earthquake records but does not seem to have had a beneficial effect on the 3 storey building in terms of the magnitude of forces transmitted to the structure.

4.2 *Equivalent period and damping*

To understand the seismic response of the building in the experiments, the equivalent SDOF period and damping ratio were calculated using an optimisation method. SDOF response was calculated at a range of period and damping values using the input motion to the structure-foundation models measured in the soil just beneath the foundation. Then the best fit to the measured structure response in the experiments was calculated by finding the minimum of the sum of the square of the residuals at each time step.

Figure 8 presents the equivalent SDOF period values that best matched the measured response for the three different building models subjected to Christchurch Earthquake records from three different strong ground motion recording stations used in the experiments (CHHC, CBGS, and CCCC). The fixed base period values for each building model are plotted for reference. An elongation of equivalent period of response to higher values than the fixed base period was observed for all of the building models and was consistent for the different input motions. This elongation is due to SFSI effects of uplift and permanent soil deformation. The shift in period was greatest for the 7 storey building and least significant for the 3 storey building.

By considering the CBGS Christchurch Earthquake record response spectra in Figure 5 in relation to the elongation of equivalent period, a reason for the different acceleration response observed for the three building models can be postulated. The amplitude of the response spectra of the Christchurch Earthquake motion remained high over a range of periods from about 0.2 s to 1 s. Therefore, for the 3 storey model, with the lowest fixed based natural period, the shift in period still meant that the structural response remained large. The 5 and 7 storey models had larger fixed base natural periods and so the shift in period of peak response to higher values coincided with the rapidly decreasing part of the input motion response spectra, which resulted in lower and attenuated structural response.

The associated equivalent SDOF damping ratios for the three building models subjected to the three Christchurch Earthquake records are plotted in Figure 9, with the assumed fixed base damping ratio of 5% plotted for reference. Equivalent damping ratios were approximately 10% for all of the building models and all of the Christchurch Earthquake records. These damping ratios represent the total damping in the system assuming a SDOF response. The higher damping ratio in the experiments than the assumed fixed base value suggests that SFSI can result in significant energy dissipation relative to that expected for a fixed base

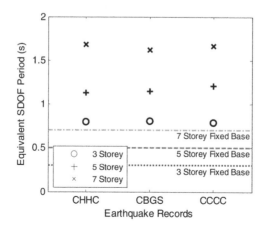

Figure 8.   Optimised equivalent SDOF period for the three building models subjected to the Christchurch Earthquake records.

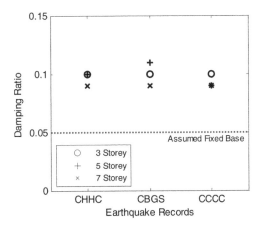

Figure 9. Optimised equivalent SDOF damping ratio for the three building models subjected to the Christchurch Earthquake records.

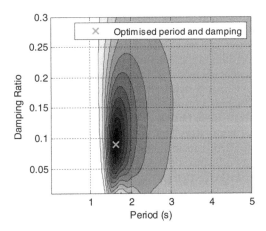

Figure 10. Contour plot of the best fit equivalent SDOF period and damping values to the measured experimental response of the 7 storey model subjected to the CBGS Christchurch Earthquake record—darker colours represent a closer match.

structure, even when the extent of uplift and permanent soil deformation is small relative to the size of the foundation.

The optimisation method allowed the results for different combinations of equivalent SDOF period and damping to be compared to investigate the accuracy with which the optimised values represented the response of the structure-foundation system. In Figure 10, a contour plot of how well different combinations of equivalent SDOF period and damping fitted the measured response (where darker colours in the plot represent a closer match) is presented for the 7 storey model subjected to the CBGS Christchurch Earthquake record. Similar results were obtained for all of the building models subjected to the Christchurch Earthquake records. It can be observed from this plot that a global best fit was obtained, which indicated that a unique combination of period and damping best fit the experimental data. It can also be seen that the contours were elongated in the direction of the damping ratio axis, suggesting that the equivalent period may be more accurately determined than the damping ratio in the analysis and a range of damping values may give a suitable fit to the measured response in the experiments.

5 CONCLUSIONS

This paper has shown that SFSI has been influential in the successful performance of multi-storey buildings on shallow foundations during the Christchurch Earthquake of February 22, 2011. Models of equivalent SDOF 3, 5, and 7 storey buildings on raft foundations were developed to represent multi-storey buildings on shallow raft foundations that performed well in the Christchurch Earthquake. Centrifuge testing of these models has shown a small extent of nonlinear interaction at the soil-foundation interface can have a significant effect on the seismic response of these structures.

Uplift of the foundation combined with permanent soil deformation was observed when the equivalent 3, 5, and 7 storey building models resting on a layer of dense sand were subjected to Christchurch Earthquake records. However, the extent of uplift and foundation settlement was not substantial compared to the overall size of the foundation, with uplift of between 0.1% and 0.5% of the width of the foundation and a maximum settlement of 10 mm for the 8 m wide foundations. These nonlinear effects influenced the acceleration response of the structures, attenuating the peak acceleration of the taller building models, which meant forces transmitted to these structures were potentially reduced.

Nonlinear SFSI resulted in increased equivalent SDOF period and a large amount of damping for all of the buildings subjected to the Christchurch Earthquake records. The elongation of period meant that the response moved away from the generally higher energy content at lower period values. This had a more significant beneficial effect on the taller building models as the shift in period coincided with the rapidly decreasing part of the input motion response spectra, which resulted in attenuated structural response. Equivalent damping was also higher than the typically assumed 5% for fixed base structures, being around 10% for all buildings subjected to the Christchurch Earthquake records. This means there is potential for significant energy dissipation as a result of SFSI and that higher damping may be assumed in equivalent SDOF analysis of these buildings if nonlinear SFSI is incorporated.

Shallow foundations of multi-story buildings in the Christchurch CBD on non-liquefying soil are likely to have uplifted, causing plastic deformation of the underlying soil, which together have significantly influenced the response of the structures during the earthquake. The small extent of the SFSI effects means there is potential for significant energy dissipation during large earthquake loading without detrimental effects on building serviceability. The experimental results showed that a rocking shallow raft foundation on a competent soil, having a large reserve of bearing capacity during uplift, could be beneficial for structural performance during earthquake loading.

REFERENCES

Anastasopoulos, I. 2010. Beyond conventional capacity design: Towards a new design philosophy. *In:* Orense, R.P., Chouw, N. & Pender, M.J. (eds.) *Soil-Foundation-Structure Interaction.* London: CRC Press/Balkema, Taylor & Francis Group.

Bertalot, D. 2013. *Seismic behaviour of shallow foundations on layered liquefiable soils.* PhD Thesis, University of Dundee.

Bolland, J. 2011. Seismic Evaluation: Club Tower 62 Worcester Boulevard Christchurch. *Buller George Turkington,*

Clifton, C. 2013. *RE: Personal communications with Luke Storie and Michael Pender regarding Apparent Strength and Stiffness of the HSBC Building with the Models in the February 22nd 2011 Earthquake.*

Cubrinovski, M., Bradley, B., Wotherspoon, L.M., Green, R., Bray, J., Wood, C., Pender, M., Allen, J., Bradshaw, A., Rix, G., Taylor, M., Robinson, K., Henderson, D., Giorgini, S., Ma, K., Winkley, A., Zupan, J., O'rourke, T., Depascale, G. & Wells, D. 2011. Geotechnical aspects of the 22 February 2011 Christchurch Earthquake. *New Zealand Society for Earthquake Engineering Bulletin,* 44(4), 205–226.

Gajan, S., Raychowdhury, P., Hutchinson, T.C., Kutter, B.L. & Stewart, J.P. 2010. Application and Validation of Practical Tools for Nonlinear Soil-Foundation Interaction Analysis. *Earthquake Spectra,* 26(1), 111–129.

Harden, C.W., Hutchinson, T.C., Martin, G.R. & Kutter, B.L. 2005. Numerical modeling of the nonlinear cyclic response of shallow foundations. *Pacific Earthquake Engineering Research Center (PEER),*

Kelly, T.E. 2009. Tentative seismic design guidelines for rocking structures. *Bulletin of the New Zealand Society for Earthquake Engineering,* 42(4), 239–274.

Loli, M., Knappett, J.A., Brown, M.J., Anastasopoulos, I. & Gazetas, G. 2014. Centrifuge modeling of rocking-isolated inelastic RC bridge piers. *Earthquake Engineering & Structural Dynamics,* 43(15), 2341–2359.

Martin, G.R. & Lam, I.P. 2000. Earthquake resistant design of foundations - Retrofit of existing foundations. *GeoEng 2000.* Melbourne, Australia.

Orense, R.P., Chouw, N. & Pender, M.J. (eds.) 2010. *Soil-Foundation-Structure Interaction,* London: CRC Press/Balkema.

Pecker, A. & Chatzigogos, C.T. 2010. Non Linear Soil Structure Interaction: Impact on the Seismic Response of Structures. *In:* Garevski, M. & Ansal, A. (eds.) *Earthquake Engineering in Europe.* Springer Netherlands.

Priestley, M.J.N., Calvi, G.M. & Kowalsky, M.J. 2007. *Displacement-based seismic design of structures,* Pavia, IUSS Press.

Schofield, A.N. (1980). Cambridge geotechnical centrifuge operations. *Géotechnique,* 30(3), 227–268.

Standards New Zealand 2004. NZS1170.5:2004 Structural Design Actions Part 5: Earthquake actions – New Zealand. Wellington, New Zealand.

Storie, L.B. 2017. *Soil-foundation-structure interaction in the earthquake performance of multi-storey buildings on shallow foundations.* PhD Thesis, University of Auckland.

Taylor, R.N. 1995. Centrifuges in modelling: priciples and scale effects. *In:* Taylor, R.N. (ed.) *Geotechnical Centrifuge Technology.* London: Blackie Academic and Professional.

Seismic Performance of Soil-Foundation-Structure Systems – Chouw, Orense & Larkin (Eds)
© 2017 Taylor & Francis Group, London, ISBN 978-1-138-06251-1

Double, triple and multiple impulses for critical elastic-plastic earthquake response analysis to near-fault and long-duration ground motions

I. Takewaki & K. Kojima
Kyoto University, Kyoto, Japan

ABSTRACT: Near-fault ground motions and long-period, long-duration ground motions possess special characteristics. The essential aspect of the near-fault ground motions can be characterized by one-cycle or a few-cycle sinusoidal waves (e.g. Ricker wavelet) which are well represented by double or triple impulses. Furthermore, the principal part of the long-period, long-duration ground motions can be characterized by many-cycle sinusoidal waves which are well described by multiple impulses. In this paper, it is shown that good approximate elastic-plastic responses of various structural models to the near-fault ground motions and long-period, long-duration ground motions can be derived by using the responses to the double, triple and multiple impulses. The energy approach plays an important and critical role in the derivation of such good approximate responses in closed form.

1 INTRODUCTION

It is well known that near-fault ground motions have peculiar characteristics and long-period, long-duration ground motions are causing great influences on high-rise buildings and base-isolated buildings. The effects of near-fault ground motions on structural response have been investigated extensively (Hall et al. 1995, Sasani and Bertero 2000, Alavi and Krawinkler 2004, Kalkan and Kunnath 2006, Khaloo et al. 2015). The fling-step and forward-directivity are two special keywords to characterize such near-fault ground motions (Mavroeidis and Papageorgiou 2003, Kalkan and Kunnath 2006). Especially, Northridge earthquake in 1994, Hyogoken-Nanbu (Kobe) earthquake in 1995 and Chi-Chi (Taiwan) earthquake in 1999 raised special attention to many earthquake structural engineers.

The fling-step and forward-directivity inputs have been characterized by two or three wavelets. For this class of ground motions, many useful research works have been conducted. Mavroeidis and Papageorgiou (2003) investigated the characteristics of this class of ground motions in detail and proposed some simple models. In this paper, several new approaches based on the double impulse (Kojima and Takewaki 2015a) and the triple impulse (Kojima and Takewaki 2015b) are proposed for various models representing important nonlinear vibration phenomena and the intrinsic response characteristics by the near-fault ground motion are captured. Then the approaches are applied to several interesting models in practice (soil-structure interaction problem, dynamic stability problem including collapse, overturning problem of rocking block). The common concept is the modeling of simple sinusoidal waves into a few impulses and the use of energy balance in the closed-form derivation of the maximum elastic-plastic response. The use of energy balance is enabled because the impulses cause only free vibration and complicated treatment by forced input can be avoided. The proposed approach is expected to overcome the difficulty of computational repetition for capturing resonant phenomena in nonlinear structural dynamics (Caughey 1960, Iwan 1961).

2 MODELING OF MAIN PART OF NEAR-FAULT GROUND MOTION INTO DOUBLE IMPULSE AND TRIPLE IMPULSE

It is known that most near-fault ground motions have a few pulse-like waves as shown in Figure 1. When only the maximum response is concerned, the response resulting from such pulse-like waves is important. In this paper, the main part of the pulse-like ground motions is modeled into the double impulse (see Figure 2(a)) or the triple impulse (see Figure 2(b)).

In this paper, the level of the double or triple impulse is adjusted so that the maximum Fourier amplitude of the double or triple impulse is equal to that of the corresponding one-cycle sine wave or one and half-cycle sine wave. This adjustment is made in order to guarantee the better response correspondence between the one-cycle sine wave or one and half-cycle sine wave and the double or triple impulse. The correspondence of the Fourier amplitudes

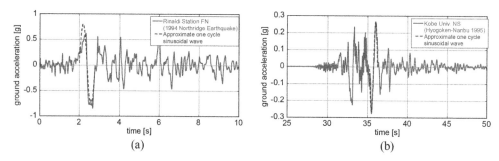

Figure 1. Modeling of main part of pulse-type recorded ground motion into the corresponding one-cycle sinusoidal input: (a) Rinaldi station fault-normal component (Northridge earthquake 1994), (b) Kobe University NS component (almost fault-normal) (Hyogoken-Nanbu (Kobe) earthquake 1995) (Kojima and Takewaki 2016a).

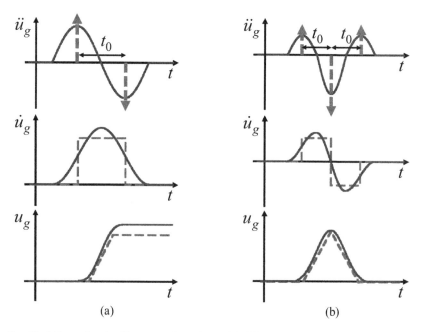

Figure 2. Modeling of pulse-like ground motions: (a) Fling-step input (one-cycle sine) and double impulse, (b) Forward-directivity input (one and half-cycle sine) and triple impulse (Kojima and Takewaki 2015a).

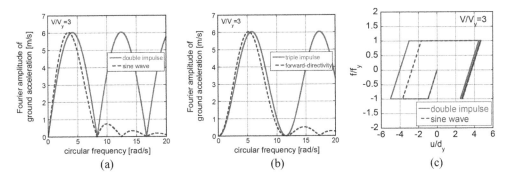

Figure 3. Adjustment of input level of double or triple impulse to the corresponding one-cycle or one and half-cycle sine wave based on Fourier amplitude equivalence and response correspondence under double impulse: (a) Double impulse, (b) Triple impulse, (c) Restoring force-deformation correspondence under double impulse (Kojima and Takewaki 2015a, b).

between the one-cycle sine wave or one and half-cycle sine wave and the double or triple impulse is shown in Figures 3(a), (b).

3 CLOSED-FORM ELASTIC-PLASTIC RESPONSE TO CRITICAL DOUBLE IMPULSE

Figure 4 shows an overview of the response process of an elastic-perfectly plastic single-degree-of-freedom (SDOF) model to the critical double impulse. The critical double impulse means the double impulse causing the maximum response under a constant velocity amplitude and a variable impulse interval (Drenicl 1970, Takewaki 2007). It should be emphasized that the critical timing of the second impulse is the time when the restoring force attains zero in the first unloading process (Kojima and Takewaki 2015a). The response correspondence (restoring force-deformation relation) under the double impulse is shown in Figure 3(c).

Figure 5 illustrates the maximum deformation for two recorded ground motions (Rinaldi station fault-normal component during Northridge earthquake 1994 and Kobe University NS component during Hyogoken-Nanbu (Kobe) earthquake 1995) and that obtained by the proposed critical double impulse. Since the recorded ground motion is fixed, the initial velocity V is fixed and V_y (product of the natural circular frequency ω_1 and the yield deformation d_y: reference velocity giving just yield deformation after the first impulse) is changed here. Because ω_1 is closely related to the resonance condition, d_y is changed principally. This procedure is similar to the well-known elastic-plastic response spectrum developed in 1960–1970 in the field of earthquake resistant design. The solid line is obtained by changing V_y for the specified V using the method for the double impulse and the dotted line is drawn by conducting the elastic-plastic time-history response analysis on each model with varied V_y under the recorded ground motion. It can be observed that the result by the proposed method is a fairly good approximation of the recorded pulse-type ground motions. If we use a bilinear hysteresis model with a positive second slope, the correspondence becomes better.

4 CLOSED-FORM ELASTIC-PLASTIC RESPONSE TO CRITICAL TRIPLE IMPULSE

In comparison with the response to the double impulse, the process of deriving the critical timing is somewhat complicated in the triple impulse. The maximum deformation after the first impulse is denoted by u_{max1}, that after the second impulse is expressed by u_{max2} and that

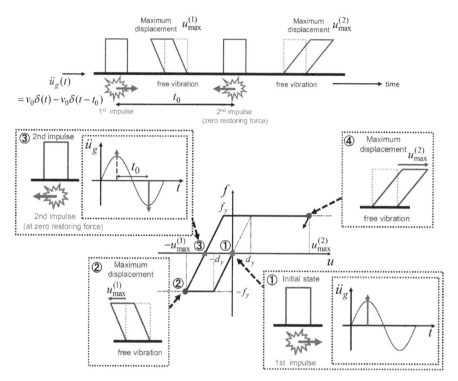

Figure 4. Overview of elastic-plastic response process of SDOF model to critical double impulse (Critical timing of second impulse is the time when the restoring force attains zero in the first unloading process.).

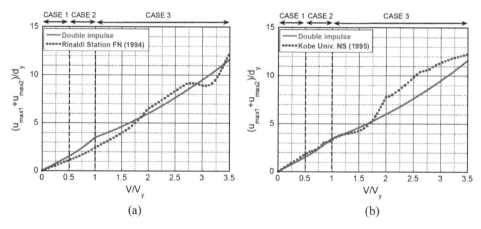

Figure 5. Maximum amplitude of deformation for the recorded ground motions and the proposed one: (a) Rinaldi station fault-normal component, (b) Kobe University NS component (Kojima and Takewaki 2016a).

after the third impulse is described by u_{max3}. The maximum deformation can be obtained by using the energy balance of kinetic energy, strain energy and dissipated energy.

Figure 6 shows the following four cases depending on the input level.

CASE 1: Elastic response during all response stages (u_{max3} is the largest)
CASE 2: Yielding after the third impulse (u_{max3} is the largest)

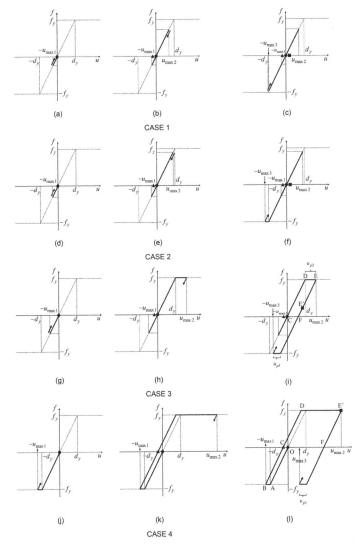

Figure 6. Prediction of maximum elastic-plastic deformation under triple impulse based on energy approach (●: first impulse, ▲: second impulse,: ■ third impulse) (Kojima and Takewaki 2015b).

CASE 3: Yielding after the second impulse (u_{max2} or u_{max3} is the largest)
    3-1: The timing of the third impulse is in the unloading stage.
    3-2: The timing of the third impulse is in the yielding (loading) stage.
CASE 4: Yielding after the first impulse (u_{max2} is the largest)

It is assumed here that the critical impulse has the second impulse timing (time of the second impulses) of zero restoring force in the first unloading process. It can be understood that the third impulse timings (time of the third impulses) are different in CASE 3 and CASE 4. Careful treatment should be made in the derivation of the critical timing (Kojima and Takewaki 2015b).

Figure 7 shows the comparison of critical ductility and earthquake input energy between the triple impulse and the forward-directivity input (one and half-cycle sine wave). It can be observed that, when the adjustment of input amplitude is made following the procedure in Section 2, the responses of the triple impulse and the forward-directivity input coincide fairly well.

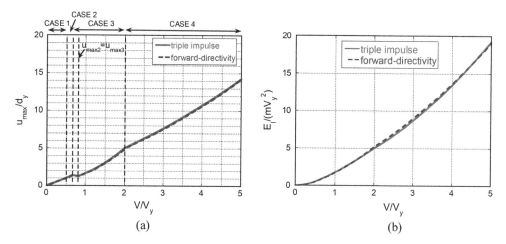

Figure 7. Comparison of triple impulse and the corresponding three wavelets of sinusoidal waves: (a) Ductility, (b) Earthquake input energy (Kojima and Takewaki 2015b).

5 CLOSED-FORM CRITICAL EARTHQUAKE RESPONSE OF ELASTIC-PLASTIC STRUCTURES ON COMPLIANT GROUND UNDER NEAR-FAULT GROUND MOTIONS

The problem of soil-structure interaction is very interesting and important in the structural and geotechnical engineering. It is aimed here that, once the soil-structure interaction model can be modeled into an SDOF model, the formulation presented in Section 3 can be applied to the soil-structure interaction model. Figure 8 presents the simplified swaying-rocking model and the equivalent SDOF model. k_e is the equivalent stiffness.

Figure 9 shows the relation of the maximum deformation ratio $(d_y + u_p)/d_y$ (d_y yield deformation, u_p plastic deformation) with V/V_y for three soil conditions and fixed-base case. In the low input level, as the ground becomes stiffer, the plastic deformation of the super-structure becomes larger. On the other hand, in the large input level, as the ground becomes softer, the plastic deformation of the super-structure becomes larger. These properties result from the fact that, as the ground becomes softer, the strain energy stored in the ground becomes larger in the case where the super-structure is in the plastic range. It is interesting to note that such properties can be derived by taking full advantage of the closed-form expression of the critical elastic-plastic responses (Kojima and Takewaki 2016a).

6 CLOSED-FORM DYNAMIC STABILITY CRITERION FOR ELASTIC-PLASTIC STRUCTURES UNDER NEAR-FAULT GROUND MOTIONS

The problem of dynamic collapse of structures has been an important and challenging problem for long time. In this section, it is demonstrated that the proposed approach (balance of input energy and dissipated energy) can be applied to this problem. Figure 10 shows several patterns of stability limit (patterns of collapse). The vertical axis is the ratio of the input velocity level to the structural strength and the horizontal axis is the second slope ratio to the first stiffness. A more detailed derivation can be found in the reference (Kojima and Takewaki 2016b).

Figure 11(a) shows the maximum deformation with respect to V/V_y under the Rinaldi station fault-normal component and the corresponding double impulse. The solid line has been drawn by the proposed method. On the other hand, the dotted line has been obtained from the time-history response analysis for many models with different values of V_y. It can

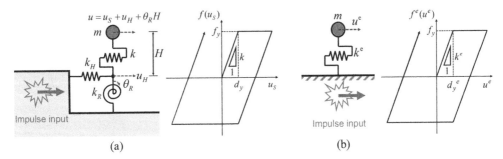

Figure 8. Modeling in soil-structure interaction problem: (a) Simplified swaying-rocking model, (b) Equivalent SDOF model (Kojima and Takewaki 2016a).

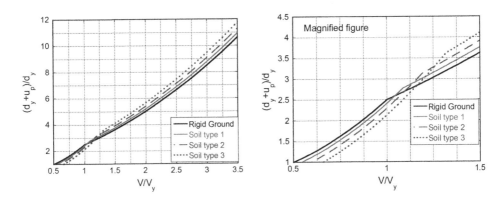

Figure 9. Relation of the maximum plastic deformation $(d_y + u_p / d_y)$ with V/V_y for three soil conditions and fixed-base case (Kojima and Takewaki 2016a).

be found that about $V/V_y = 0.8$ is the approximate limit. From the detailed investigation, $V/V_y = 0.78$ and $V/V_y = 0.79$ have been selected for candidates to be investigated. Figure 11(b) demonstrates the restoring-force-deformation relation for the stable case ($V/V_y = 0.78$) and the unstable case ($V/V_y = 0.79$). In addition, Figure 11(c) presents the deformation time-history for the stable case ($V/V_y = 0.78$) and the unstable case ($V/V_y = 0.79$). On the other hand, Figure 11(d) shows the corresponding restoring-force time-history for stable case ($V/V_y = 0.78$) and the unstable case ($V/V_y = 0.79$). It can be confirmed that the proposed stability limit using the double impulse is fairly accurate.

7 CLOSED-FORM OVERTURNING LIMIT OF RIGID BLOCK UNDER CRITICAL NEAR-FAULT GROUND MOTIONS

A closed-form limit on the input level of the double impulse as a substitute of a near-fault ground motion can be derived for the overturning of a rigid block (Nabeshima et al. 2016). Figure 12 shows the modeling of the rocking motion of a rigid block using a rigid bar supported by a non-linear elastic rotational spring with rigid initial stiffness and negative second slope.

Figure 13 presents the rocking response of a rigid block and governing law (conservation of angular momentum, conservation of energy, energy dissipation).

Figure 14 illustrates the limit velocity amplitude of the critical double impulse with respect to R for $2b = 1, 2, 4[m]$ (closed-form expression and numerical simulation). The proposed closed-form limit velocity amplitude coincides fairly well with the numerical simulation result. Figure 14 demonstrates that the proposed method seems reliable.

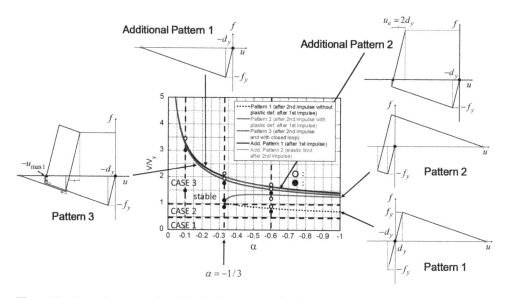

Figure 10. Several patterns of stability limit (patterns of collapse) (Kojima and Takewaki 2016b).

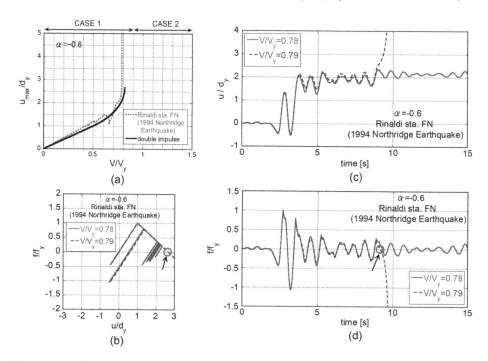

Figure 11. Stable model ($V/V_y = 0.8$) and unstable model ($V/V_y = 0.79$) under Rinaldi station fault-normal component: (a) Maximum deformation with respect to V/V_y under Rinaldi station fault-normal component and the corresponding double impulse, (b) Restoring force-deformation relation for stable case and unstable case, (c) Deformation time-history for stable case and unstable case, (d) Restoring-force time-history for stable case and unstable case (Kojima and Takewaki 2016b).

Figure 15 shows the critical acceleration amplitude ratio of the equivalent one-cycle sinusoidal input to acceleration of gravity for $2b = 1, 2, 4$[m] and the comparison with other results (Dimitrakopoulos and DeJong 2012, Makris and Kampas 2016). The accuracy of the proposed method can be assured.

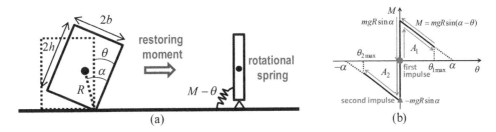

(a)   (b)

Figure 12. Modeling of rocking rigid block: (a) Modeling by rigid bar supported by non-linear elastic rotational spring with rigid initial stiffness and negative second slope, (b) Moment-rotation relation for rocking response of rigid block and timing of double impulse.

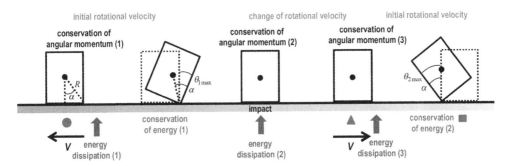

Figure 13. Rocking response of rigid block and governing law (conservation of angular momentum, conservation of energy, energy dissipation) (Nabeshima et al. 2016).

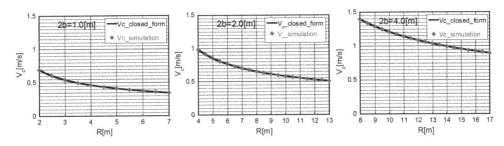

Figure 14. Limit velocity amplitude of critical double impulse with respect to R for $2b = 1, 2, 4$[m] (closed-form expression and numerical simulation).

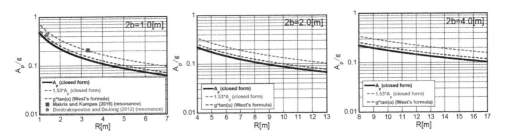

Figure 15. Critical acceleration amplitude ratio of equivalent one-cycle sinusoidal input to acceleration of gravity for $2b = 1, 2, 4$[m] and comparison with other results (Nabeshima et al. 2016).

8 SIMPLE EVALUATION METHOD OF SEISMIC RESISTANCE OF RESIDENTIAL HOUSE UNDER TWO CONSECUTIVE SEVERE GROUND MOTIONS

In the 2016 Kumamoto earthquake in Japan, two severe ground shakings with the seismic intensity 7 (the highest level in Japan Metheorological Agency (JMA) scale; approximately X-XII in Mercalli scale) occurred consecutively on April 14 and April 16. In the seismic regulations of most countries, it is usually prescribed that such severe earthquake ground motion occurs once in the working period of buildings. A simple evaluation method has been proposed on the seismic resistance of residential houses under two consecutive severe ground motions with intensity 7 (Kojima and Takewaki 2016c). In that paper, an impulse of the velocity V has been adopted as a representative of near-fault ground motion and two separated impulses have been used as the repetition of intensive ground shakings with the seismic intensity 7 (see Figure 16).

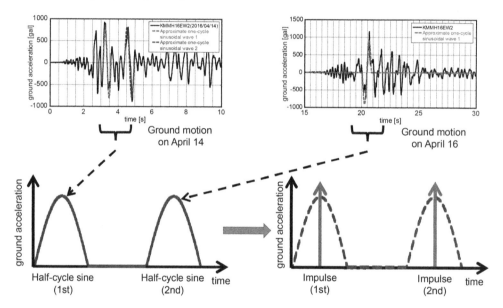

Figure 16. Modeling of repeated intensive ground motions into two impulses (Kojima and Takewaki 2016c).

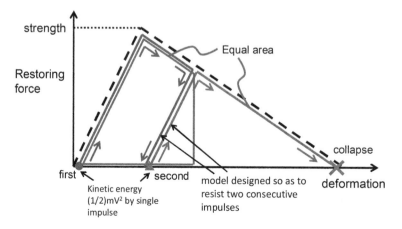

Figure 17. Collapse scenario under two impulses and energy consideration for evaluating limit input level (Kojima and Takewaki 2016c).

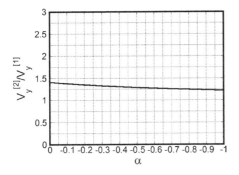

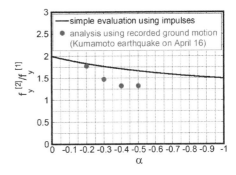

Figure 18. $V_y^{[2]}/V_y^{[1]}$ for α (Kojima and Takewaki 2016c).

Figure 19. $f_y^{[2]}/f_y^{[1]}$ for α (Kojima and Takewaki 2016c).

Figure 17 presents the collapse scenario under two impulses and energy consideration for evaluating limit input velocity. Figure 18 illustrates the $V_y^{[2]}/V_y^{[1]}$ for the second-slope ratio α where $V_y^{[1]}$ denotes the reference velocity (strength of the model) just collapsing after one impulse and $V_y^{[2]}$ denotes the reference velocity (strength of the model) just collapsing after two impulses. Finally the plot of $f_y^{[2]}/f_y^{[1]}$ (the ratio of the model strength for just collapse after two impulses to that for just collapse after one impulse) is shown in Figure 19. The simulation using the Kumamoto earthquake ground motion (on April 16) is also included.

9 CLOSED-FORM ELASTIC-PLASTIC RESPONSE TO CRITICAL MULTIPLE IMPULSE

Long-period, long-duration ground motions are of great concern recently after the Mexico earthquake in 1985, the Tokachioki earthquake in 2003 and the Tohoku (Japan) earthquake in 2011 (Kojima and Takewaki 2015c). Figure 20 shows an actual resonant response of a super high-rise building in Osaka, Japan during the 2011 off the Pacific coast of Tohoku earthquake. This phenomenon clearly indicates the necessity and requirement of consideration of response under long-duration ground motion.

The multiple impulse input can be used as a substitute of the long-duration earthquake ground motion, mostly expressed in terms of harmonic waves, and a closed-form solution can be derived of the elastic-plastic response of an SDOF structure under the critical multiple impulse input. While the critical set of input amplitude and input frequency (timing of impulse) have to be computed iteratively for the multi-cycle sinusoidal wave, that can be obtained directly without iteration for the multiple impulse input by introducing a modified version (only the timing between the first and second impulses is modified so that the second impulse is given at the zero restoring-force). The resonance can be proved by using energy investigation. The critical timing of the multiple impulses can be characterized as the time with zero restoring force. This decomposition of input amplitude and input frequency has overcome the long-time difficulty in finding the resonant frequency without repetition.

Since only the free-vibration appears in such multiple impulse input, the energy approach plays an important role in the derivation of the closed-form solution of a complicated elastic-plastic critical response. In other words the energy approach enables the derivation of the maximum critical elastic-plastic seismic response without solving the differential equation (equation of motion). In this process, the input of impulse is expressed by the instantaneous change of velocity of the structural mass. The maximum elastic-plastic response after impulse can be obtained by equating the initial kinetic energy computed by the initial velocity to the sum of hysteretic and elastic strain energies as in the formulation under the double and triple impulses.

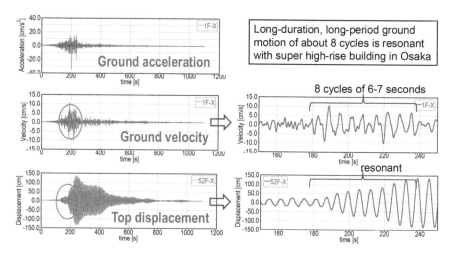

Figure 20. Resonant response of a super high-rise building in Osaka, Japan during the 2011 Tohoku (Japan) earthquake under long-duration, long-period ground motion (Kojima and Takewaki 2015c).

10 CONCLUSIONS

In this paper, it has been shown that good approximate elastic-plastic responses of various structural models to near-fault ground motions and long-duration ground motions can be derived by using the responses to the corresponding double, triple and multiple impulses. The original energy approach played an important role in the derivation of such good approximate responses in closed form. It should be emphasized that even the phenomena expressed by the negative second slope can be treated in a unified manner.

ACKNOWLEDGEMENTS

Part of the present work is supported by the Grant-in-Aid for Scientific Research (KAKENHI) of Japan Society for the Promotion of Science (No.15H04079). This support is greatly appreciated. A recorded ground motion was provided by KiK-net.

REFERENCES

Alavi, B. and Krawinkler, H. 2004. Behaviour of moment resisting frame structures subjected to near-fault ground motions, *Earthquake Eng. Struct. Dyn.* **33**(6), 687–706.

Caughey, TK. 1960. Sinusoidal excitation of a system with bilinear hysteresis. *J. Appl. Mech.* **27**(4), 640–643.

Dimitrakopoulos, E. G. and DeJong, M. J. 2012. Revisiting the rocking block: closed-form solutions and similarity laws, *Proc. R. Soc. A*, **468**, 2294–2318.

Drenick, RF. 1970. Model-free design of aseismic structures. *J. Eng. Mech. Div.*, ASCE, **96**(EM4), 483–493.

Hall, J. F., Heaton, T. H., Halling, M. W., and Wald, D. J. 1995. Near-source ground motion and its effects on flexible buildings, *Earthuake Spectra*, **11**(4), 569–605.

Iwan, W. D. 1961. *The dynamic response of bilinear hysteretic systems*, Ph.D. Thesis, California Institute of Technology.

Kalkan, E. and Kunnath, S.K. 2006. Effects of fling step and forward directivity on seismic response of buildings, *Earthquake Spectra*, **22**(2), 367–390.

Khaloo, A.R., Khosravi1, H. and Hamidi Jamnani, H. 2015. Nonlinear interstory drift contours for idealized forward directivity pulses using "Modified Fish-Bone" models; *Advances in Structural Eng.***18**(5), 603–627.

Kojima, K. and Takewaki, I. 2015a. Critical earthquake response of elastic-plastic structures under near-fault ground motions (Part 1: Fling-step input), *Frontiers in Built Environment* (Specialty Section: Earthquake Engineering), Volume 1, Article 12.

Kojima, K. and Takewaki, I. 2015b. Critical earthquake response of elastic-plastic structures under near-fault ground motions (Part 2: Forward-directivity input), *Frontiers in Built Environment* (Specialty Section: Earthquake Engineering), Volume 1, Article 13.

Kojima, K. and Takewaki, I. 2015c. Critical input and response of elastic-plastic structures under long-duration earthquake ground motions, *Frontiers in Built Environment* (Specialty Section: Earthquake Engineering), Volume 1, Article 15.

Kojima, K. and Takewaki, I. 2016a. Closed-form critical earthquake response of elastic-plastic structures on compliant ground under near-fault ground motions, *Frontiers in Built Environment* (Specialty Section: Earthquake Engineering), Volume 2, Article 1.

Kojima, K. and Takewaki, I. 2016b. Closed-form dynamic stability criterion for elastic-plastic structures under near-fault ground motions, *Frontiers in Built Environment* (Specialty Section: Earthquake Engineering), Volume 2, Article 6.

Kojima, K. and Takewaki, I. 2016c. A simple evaluation method of seismic resistance of residential house under two consecutive severe ground motions with intensity 7, *Frontiers in Built Environment* (Specialty Section: Earthquake Engineering), Volume 2, Article 15.

Makris, N and Kampas, G. 2016. Size versus slenderness: Two competing parameters in the seismic stability of free-standing rocking columns, *Bull. Seismol. Soc. Am.*, **106**(1) published online.

Mavroeidis, G. P., and Papageorgiou, A. S. 2003. A mathematical representation of near-fault ground motions, *Bull. Seism. Soc. Am.*, **93**(3), 1099–1131.

Nabeshima, K., Taniguchi, R., Kojima, K. and Takewaki, I. 2016. Closed-form overturning limit of rigid block under critical near-fault ground motions, *Frontiers in Built Environment* (Specialty Section: Earthquake Engineering), Volume 2, Article 9.

Sasani, M. and Bertero, V.V. 2000. "Importance of severe pulse-type ground motions in performance-based engineering: historical and critical review," in *Proceedings of the Twelfth World Conference on Earthquake Engineering*, Auckland, New Zealand.

Takewaki, I. 2007. *Critical excitation methods in earthquake engineering*, Elsevier, Second edition in 2013.

Taniguchi, R., Kojima, K. and Takewaki, I. 2016. Critical response of 2DOF elastic-plastic building structures under double impulse as substitute of near-fault ground motion, *Frontiers in Built Environment* (Specialty Section: Earthquake Engineering), Volume 2, Article 2.

Seismic Performance of Soil-Foundation-Structure Systems – Chouw, Orense & Larkin (Eds)
© 2017 Taylor & Francis Group, London, ISBN 978-1-138-06251-1

Performance based design of a structural foundation on liquefiable ground

A.K. Murashev
Opus, FIPENZ, Wellington, New Zealand

C. Keepa
Opus, MIPENZ, Wellington, New Zealand

A. Tai
Opus, Auckland, New Zealand

ABSTRACT: Seismic behaviour of buildings on liquefiable ground is affected by the size and stiffness of the structural foundation, level of contact pressure, seismic response of the structure and soil, thickness and properties of liquefiable soil layers and non-liquefiable crust, intensity of ground motion and many other factors. Costly ground improvement is commonly carried out to stabilise liquefiable soils. In many cases seismic performance requirements for buildings and structures located on liquefiable sites with a non-liquefiable crust can be satisfied without ground improvement. A foundation system comprising a Reinforced Concrete (RC) raft foundation over liquefiable soils with non-liquefiable crust was designed to support the recently completed Rotorua Police Station building. The adopted design framework included dynamic time-history finite element analysis of soil-foundation-superstructure interaction. Performance based design utilised in the analysis of dynamic soil-foundation-superstructure interaction resulted in substantial cost savings due to avoidance of ground improvement.

1 INTRODUCTION

Seismic behaviour of buildings on liquefiable ground is affected by the size and stiffness of the structural foundation, magnitude of contact pressure, seismic response of the structure and soil, thickness and properties of liquefiable soil layers and non-liquefiable crust, intensity of ground motion and many other factors. While it is common for geotechnical engineers to assess liquefaction-induced settlement of buildings based on free field settlement due to liquefaction (when only volumetric deformation is considered), this approach is deficient and can lead to deficient designs. The actual mechanism of settlement associated with liquefaction is much more complex. The effect of structural inertia forces and shear deformation of the liquefied ground on the magnitude of settlement can be very substantial in many cases. In addition to the settlement associated with volumetric strain, buildings can experience settlement associated with shear deformation or deviatoric strain. This effect is more prominent for tall buildings with high contact pressures as such structures generate high shear stresses in the liquefied soil during seismic shaking. According to Bray & Dashti (2013), liquefaction-induced settlement of buildings is affected by the following displacement mechanisms (Figure 1):

a. Volumetric strains caused by water flow in response to transient gradients
b. Partial bearing failure due to soil softening
c. Seepage erosion of soils beneath footings with the dissipation of excess porewater pressures
d. Soil-structure-interaction-induced building ratcheting during earthquake loading

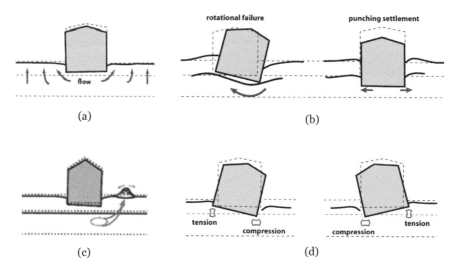

Figure 1. Liquefaction-induced mechanisms affecting settlement (Bray & Dashti, 2013) a) Volumetric strains (b) Partial bearing failure due to soil softening (c) Seepage erosion (d) Soil-structure-interaction-induced building ratcheting during earthquake loading.

While the mechanism of seismic settlement of structures on liquefiable ground is still a subject of extensive research, the available case studies from previous earthquakes as well as the existing empirical and analytical design procedures can be used to design foundation systems at potentially liquefiable sites. In cases where there is a thick enough non-liquefiable crust, it is possible to completely avoid ground improvement and utilise the potential benefits of lower foundation stiffness and greater foundation damping in the design of the structure. Recent research work by Karamitros (2013) indicated that it is possible to ensure satisfactory foundation performance in terms of acceptable settlements and adequate post-seismic static bearing capacity, in the presence of a reasonably thick and shear-resistant non-liquefiable soil crust. The exceptions are sites that could be susceptible to lateral spreading or where there is a large contrast in ground stiffness or liquefaction potential across the site.

The early stage concept design for the Rotorua Police building required several million dollars of ground improvement to mitigate effects of liquefaction. Through detailed analysis, it was shown that the natural non-liquefiable crust was of sufficient depth and consistency to mitigate the effects of liquefaction at depth without the need for ground improvement. This paper describes a design framework used for the design of the Rotorua Police building founded on liquefiable ground with a non-liquefiable soil crust. The design framework utilises available empirical, analytical and numerical methods. The design procedures used by the authors on the Rotorua Police project resulted in substantial cost savings due to avoidance of ground improvement.

2 STRUCTURES ON LIQUEFIABLE GROUND

Based on observations after past earthquakes, Ishii and Tokimatsu (1988) concluded that for buildings where the width of the foundation is large (2–3 times thickness of the liquefiable soil layer), the settlement of the building is approximately equal to the free-field settlement of the ground surface (i.e. settlement of the ground surface is not affected by structural loads). According to Seed et al. (2001), punching/bearing settlements can be expected to be large (many tens of centimetres or more) when post-liquefaction strength provide a factor of safety of less than 1.0 under gravity loading (without additional vertical loads associated with earthquake-induced rocking). These types of punching/bearing settlements can be expected to be small (less than 3 to 5 cm) when liquefaction occurs but the minimum factor

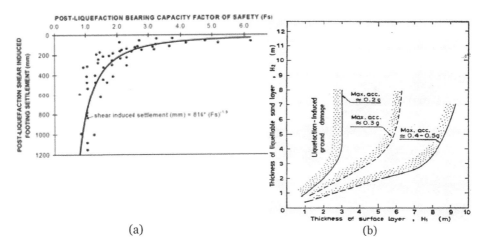

(a)   (b)

Figure 2.   Correlation between post-liquefaction shear-induced settlement and post-liquefaction bearing capacity factor of safety, Naesgaard et al., 1998 (a); correlations for the assessment of liquefaction-induced damage, Ishihara, 1985 (b).

of safety under the worst case combination of seismically-induced transient vertical loads plus static gravity loads and based on post-liquefaction strength is greater than about 2.0.

For light structures (1 to 3 storey buildings) founded on a non-liquefiable cohesive crust over liquefied soil, settlement associated with the shear strain deformations can be estimated using Figure 2a developed by Naesgaard et al. (1998) based on limited total-stress dynamic numerical analyses. The shear strain induced settlement is additional to the post-liquefaction free field consolidation settlements.

Ishihara (1985) developed correlations between surface manifestation of liquefaction (such as surface rupture and sand boils) and thicknesses of the liquefied layer and the overlying non-liquefied crust, as shown in Figure 2b. The liquefaction manifestation correlations shown in Figure 2b can be considered for light (one to three storey) buildings.

Other simplified design procedures include (Andersen et al., 2007):

– Setting the foundation punching resistance such that the shear strength of the non-liquefiable crust around the perimeter of the foundation is greater than the foundation load, or
– Setting the foundation punching resistance such that the shear strength of the non-liquefiable crust around the perimeter of the foundation plus the bearing capacity of the underlying liquefied soil using the residual strength is greater than the foundation load.

An important point of note that is borne out from the studies cited above is that the presence of a liquefiable layer overlain by a non-liquefiable crust of sufficient thickness and shear strength would not necessarily result in the manifestation of adverse effects at the ground surface even when the layer has undergone liquefaction after an earthquake. A possible explanation for this effect is that for the case of shallow foundation the majority of shear deformations due to the application of the foundation loads are expected to concentrate in the vicinity of the foundation. It is noted that this explanation is consistent with the results of the analysis reported in this paper (for example, see Figures 3 and 4).

Alternatively, numerical methods can also be used to design foundation systems for buildings founded on liquefiable ground. Numerical (e.g. finite element or finite difference) models can describe complex soil profiles and soil—structure interaction effects. Total stress and effective stress analysis procedures are available to assess triggering of liquefaction and the effects of liquefaction on the seismic response of the foundation-structure system. 2D and 3D numerical modelling has been carried out by a number of researchers and, while not free from limitations, has been proven to provide good quality results in many cases

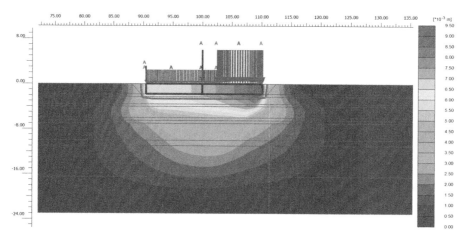

Figure 3. Vertical displacements for the static load case.

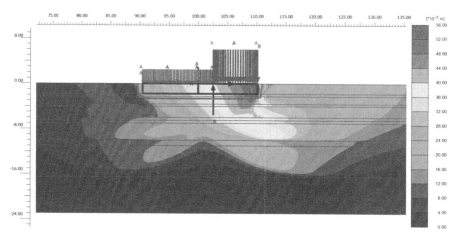

Figure 4. Vertical displacements from the seismic pushover analysis for the cyclic liquefaction phase.

(Bray & Dashti, 2012). Earthquake time histories are required as input ground motions in these analyses. Both simplified and numerical design procedures described above have been used by the authors to design a foundation system for the new Rotorua Police building.

3 SITE CONDITIONS

The recently completed three storey Rotorua Police Station building (plan size 26 m × 80 m) has been designed as an Importance Level 4 structure, serving post-disaster function and is to remain fully functional as critical infrastructure following a disaster such as a major seismic event. The structural design incorporates low damage self-centring PREcast Seismic Structural Systems technology adopted as an alternative to a more conventional reinforced concrete (RC) frame structure. Post-tensioned steel tendons were used to clamp rocking concrete shear walls to the foundations to resist earthquake forces, while remaining very stiff under serviceability loadings. The building is founded on a stiff RC raft foundation over potentially liquefiable ground with non-liquefiable crust. Close interaction between the geotechnical designers (Opus) and the structural designers (Spiire) was required to develop a cost-effective design of the foundation system for the building. For buildings of high importance, the

New Zealand Building Code requires buildings be designed for three damage states, two serviceability limit states, SLS 1 and SLS 2 and the ultimate limit state, ULS. At SLS 1 damage should be minimal and easily repairable. At SLS 2 some damage is acceptable but there should be no loss of function. For the ultimate limit state, damage is acceptable but it should not result in loss of life. The design peak ground accelerations are as follows: SLS2 PGA = 0.27 g; ULS PGA = 0.48 g.

The site is formed by undifferentiated pumiceous alluvium and is a geothermal site with low geothermal activity. The ground water table was 5 m below the ground surface level at the time of the investigations, but was conservatively assumed to be at 4 m depth for the design. Geotechnical investigations for the project comprised boreholes, seismic cone penetration tests (CPTs) with shear wave velocity measurements, grading and Atterberg Limit tests. The site ground conditions and extent of liquefaction based on the analysis of CPT data are summarised in Table 1.

Available methods for the assessment of liquefaction potential are based on case studies for hard-grained alluvial or marine soils and their applicability to pumiceous soils has not been verified. Well recorded and analysed liquefaction case studies for pumiceous sands are very limited. Calibration chamber test results reported by Wesley et al (1998) indicate that end cone resistance recorded for dense pumice sands are similar to those measured for loose pumice sands, with relatively low values usually being measured. This means that CPT results may not necessarily provide an accurate representation of the relative density of a pumice sand, and that its use for assessing the liquefaction potential on pumice sands could thus lead to overly conservative results. Recent studies by Orense et al. (2012) and Orense and Pender (2013) have further corroborated this finding, stating that the use of conventional methods for the assessment of liquefaction potential based on CPT data results in a conservative estimate of liquefaction potential for pumiceous soils and that liquefaction assessment based on Shear Wave Velocity test data may be more reliable.

In the liquefaction potential assessment carried out for the Rotorua Police Station project, both the CPT and shear wave velocity results were taken into account as per the methods described by Idriss and Bolanger (2008) and Kayen (2013). Nevertheless, as per the discussions above, there is greater confidence with the assessment using the shear wave velocity results due to the pumiceous nature of the materials found at the site. The measured shear wave velocities for the site range from 300 m/s to 1300 m/s, which indicates that the site soils have low potential for liquefaction (Youd & Idriss, 2001).

Results of cyclic triaxial tests reported by Orense et al (2012) show that alluvial pumice deposits, similar to those found at the Rotorua Police Station site, tend to possess cyclic resistance to liquefaction that is greater than other sands of comparable relative density. We note that Orense et al's test results are consistent with the results of our recent cyclic triaxial tests conducted for the Waikato Expressway Project, which indicate that cyclic resistance ratio of pumiceous sands and silts is 50–100% higher than that for hard-grained (quartz) sands (Jacobs & Dennison, 2015). Notwithstanding these findings, it was assumed for the geotechnical design that the site soils would behave in accordance with the liquefaction assessment based on the CPT data due to the unreliability of available methods of assessment of

Table 1. Summary of the site ground conditions.

Unit no	Depth range, m	Soil description	Liquefaction for SLS2 & ULS events—depth range, m
1	0 to 4	Interbedded pumiceous sandy gravel and silty sand	No liquefaction (above GWT)
2	4 to 7	Pumiceous silt	5.5–5.8; 6.5–6.8 (based on CPT)
3	7 to 11	Pumiceous silty sand with minor gravel	No liquefaction
4	11 to 16.5	Pumiceous sand	11–16.5 (based on CPT)
5	16.5 to 25	Pumiceous silt and silty sand	Localised liquefaction in up to 0.3 m think layers (based on CPT)

liquefaction potential for pumiceous soils. This assumption is conservative as the measured high shear wave velocities indicate that the site soils are non-liquefiable.

4 GEOTECHNICAL ANALYSIS FRAMEWORK

The following geotechnical analysis framework has been developed and used:

1. Assessment of the liquefaction potential of the soils is undertaken based on borehole, CPT, shear wave velocity and laboratory test data. This includes developing a ground model for the site, assessing the susceptibility of each soil layer based on their measured plasticity index and the soil descriptions, assessing whether liquefaction will be triggered using a conventional empirical approach (eg Boulanger and Idriss, 2014) and calculating free field subsidence and other liquefaction performance metrics such as the Liquefaction Severity Number (LSN).
2. Static bearing capacity and settlement are assessed and compared against design requirements. Bearing capacity is assessed using the verification method B1/VM4 in the New Zealand Building Code. Static settlement (under gravity loads) can be assessed using well established methods and numerical analysis. For the Rotorua Police Station project a finite element model of the raft—soil system was developed using the computer program Plaxis 2D and used to assess static settlement;
3. Seismic bearing capacity is assessed for three phases, an excess pore pressure develoment phase with the full pseudo-static inertia of the superstructure, a post liquefaction cyclic phase where the soils are liquefied with somewhat reduced structure inertia and a post liquefaction phase with the residual liquefied strength estimated from common empirical relationships. Stresses in the non-liquefied crust and the underlying liquefied soil layer are estimated from pseudo-static finite element analysis with reduced stiffnesses and strength for liquefied soils for each phase;
4. The effects of liquefaction on the shallow building foundation are considered using the simplified methods described in Section 2 of this paper;
5. Resistance to sliding of the structure under seismic load is assessed recognising the transitory nature of earthquake loading;
6. Analysis of the stress-strain state of the raft foundation is undertaken based on modelling of the raft on bi-linear Winkler springs allowing for differential ground displacements or loss of support to areas of the raft. A 3D model of the raft foundation was developed for the Rotorua Police Station and analysed using computer program SAFE. Static and dynamic analyses of the raft are carried out; and
7. Dynamic time history finite element analysis of the soil-foundation-structure interaction is undertaken (for the Rotorua Police Station project Plaxis 2D was used).

5 GEOTECHNICAL CONSIDERATIONS

The framework described above was applied in the design of the foundations for the Rotorua Police station. From this analysis, the expected static settlement of the raft foundation, founded 1.8 m below ground surface level, is expected to be in the order of 10 mm. Under a SLS1 event the site soils are not susceptible to liquefaction, and therefore no subsidence associated with liquefaction is expected for the SLS1 event. Subsidence of liquefied ground (or free field settlement) associated with densification of liquefiable soils is expected to vary from 50 mm for the SLS2 event to 120 mm for the ULS event. The expected differential settlement associated with the subsidence (based on the free field analysis of soil profiles encountered by different CPTs) is in the order of 10 to 30 mm.

According to Ishii and Tokimatsu (1988), for the large size of the foundation footprint of the Rotorua Police Station (26 m × 80 m), the seismic settlement of the building should be approximately equal to the free-field settlement of the ground surface (90–120 mm for a ULS event). Assessment of the bearing capacity indicated that for the post-earthquake phase with residual liquefied soil strength the factor of safety against bearing capacity failure was 2.6. Therefore, according to Seed et al. (2001), punching/bearing settlements can be expected to be small (less than 3 to 5 cm). According to the graph developed by Naesgaard et al. (1998) and shown on Figure 2a, the expected seismic settlement of the building from shear within the liquefied soil layer in the ULS event is 130 mm. Also, the thickness of the non-liquefiable crust according to Ishihara (1985) is of sufficient thickness to prevent surface manifestation of liquefaction. Vertical displacements of soil from the 2D finite element analysis (static and seismic pushover analyses) are shown on Figures 3 and 4. The raft was assumed to be rigid in these analyses.

A foundation system comprising a RC raft (with a grid of RC walls), similar to a cellular raft, as shown in Figure 5 was adopted by the geotechnical and structural designers. The structural designers analysed the stress—strain state of the raft using a Winkler model with bi-linear soil springs reflecting the ground stiffness and strength for each phase. Also, in this analysis, soil springs where removed within 4×4 m zones at various locations beneath the raft to model the effect of differential settlement of soils for the post-earthquake phase.

A more complex dynamic finite element time history analysis of the soil-foundation-structure interaction using Plaxis 2D was carried out to assess the effect of dynamic ratcheting (see Figure 1c). The time histories from four historical earthquakes records from stiff soil sites, adjusted to represent the site seismic hazard were used in the analysis. The superstructure was modelled as a two-storey portal frame constructed of plate elements. Node-to-node anchors were added in the form of diagonal cross bracings to provide the large lateral stiffness of the shear walls. The ground floor and foundation were modelled as a thick slab consisting of a soil cluster enclosed within four plate elements, where the plate element on the bottom represents the raft foundation. The mass assigned to the plate element on each floor was back-calculated from the equivalent static loads given by the structural engineers. This has been done to account for the inertia forces acting on the structure in the time-history analyses.

Interface elements were used to simulate the soil-foundation interaction on the base of the raft foundation. Two material models were used for the dynamic analysis: the Mohr-Coulomb model (M-C) and the Hardening Soil model with small-strain stiffness (HSsmall). The stiffness values of the soils used in the model were derived from CPT results. Liquefied soils were modelled as M-C materials where the stiffness values were reduced incrementally from their pre-liquefied values and the shear strength was reduced to the residual shear strength. The drainage type for all the materials used in this Plaxis model was assumed to be undrained, where calculations for the non-liquefiable and liquefied soil layers were based on effective and total stress values respectively. While it is acknowledged that the modelling methodologies described hitherto constitute a relatively simplified approach, it is anticipated that there are sufficient details present in the model to provide for a satisfactory simulation on how the foundation system would likely behave when part of the underlying soils have suddenly undergone a drastic reduction in stiffness and strength, as is the case when liquefaction occurs in the soil.

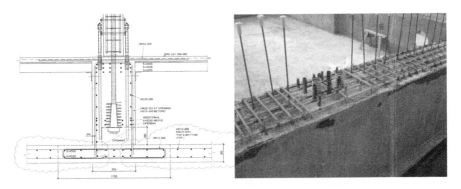

Figure 5.   Typical detail of the adopted RC raft foundation.

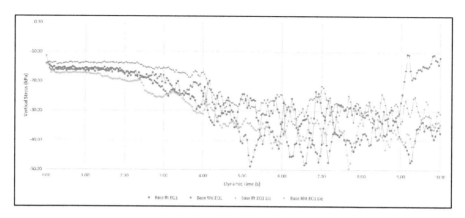

Figure 6. Vertical stress at the corners of the raft foundation for 1989 Loma Prieta earthquake time history: orange & blue—no liquefaction; yellow & pale—with liquefaction.

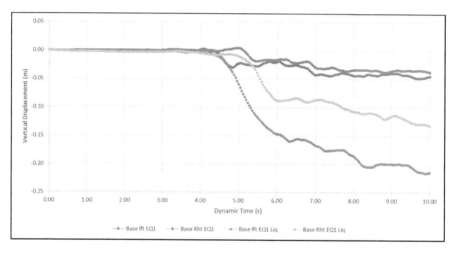

Figure 7. Settlement at the corners of the raft foundation for 1989 Loma Prieta earthquake time history: orange & blue—no liquefaction; yellow & pale—with liquefaction.

Typical results of the time-history analysis are shown in Figures 6 and 7. Our analysis indicates that, assuming that the site soils will liquefy, ratcheting may result in the maximum total settlement of approximately 200 mm and differential settlement of 70 mm over the 26 m width of the raft, which is acceptable for the ULS event (collapse avoidance).

6 CONCLUSIONS

A foundation system comprising a RC raft foundation over liquefiable soils with a sufficiently thick non-liquefiable crust was designed to support the recently constructed Rotorua Police Station building. A framework for the design of foundations has been developed for performance based design utilizing simplified approaches and less common dynamic finite element analysis. The adopted design framework included conservative assumptions with respect to the site soils' potential for liquefaction and dynamic time-history finite element analysis of soil-foundation-superstructure interaction. Implementation of the design framework in the design of foundations for the Rotorua Police station resulted in substantial cost savings due to avoidance of ground improvement.

ACKNOWLEDGMENTS

Spiire New Zealand Ltd was the structural designer for the project. RDT Pacific is gratefully acknowledged for their project management input through the course of the geotechnical and structural design work.

REFERENCES

Anderson, D.L., et al. Geotechnical Design Guidelines for Buildings on Liquefiable Sites in accordance with NBC 2005 for Greater Vancouver Region, Vancouver Liquefaction Task Force Report, 2007.

Bray, J.D., Dashti, S. Liquefaction-Induced Building Movements, Proceedings of the 2nd International Conference on Performance-Based Design Earthquake Geotechnical Engineering, Taormina, Italy, 2012.

Bray, J.D., Dashti, S. Liquefaction-Induced Movements of Buildings with Shallow Foundations, 5th International Conference on Recent Advances in Geotechnical Earthquake Engineering and Soil Dynamics, San Diego, 2010.

Dashti, S., Bray, J.D. 2013. Numerical Simulation of Building Response on Liquefiable Sand. J. Geotechnical and Environmental Engineering, 139(8), 1235–1249.

Idriss I.M., and Boulanger R.W. Soil liquefaction during earthquakes, Earthquake Engineering Research Institute MNO-12, 2008.

Ishihara, K. Stability of natural deposits during earthquakes, Proceedings of 11th International Conference on Soil Mechanics and Foundation Engineering, A.A. Balkema Publishers, Rotterdam, Netherlands, 1985.

Ishii, Y., Tokimatsu, K. Simplified procedures for the evaluation of settlements of structures during earthquakes, Proceedings from the Ninth World Conference on Earthquake Engineering, Tokyo-Kyoto, Japan, 1988.

Jacobs E. and Dennison D. Liquefaction Resistance of Hamilton Pumicious Sands, Opus Report G3336, 2015.

Karamitros D.K., Bouckovalas G.D. and Chaloulos Y.K. "Insight into the Seismic Liquefaction Performance of Shallow Foundations", ASCE Journal of Geotechnical and Geoenvironmental Engineering, vol. 139(4), 2013.

Kayen R. Shear-wave velocity based probabilistic and deterministic assessment of seismic liquefaction, Journal of Geotechnical and Geoenvironmental Engineering, Volume 139, Issue 3, 2013.

Naesgaard, E., Byrne, P.M. and Ven Huizen, G. Behaviour of light structures founded on soil crust over liquefied ground., Proc. Geotechnical Earthquake Engineering & Soil Dynamics III, ASCE Geotechnical Special Publication No. 75, 1998.

Seed, R.B., et al. Recent Advances in Liquefaction Engineering and Seismic Site Response Evaluation, Proc. of 4th International Conference on Recent Advances in Geotechnical Engineering and Soil Dynamics, San Diego, 2001.

Orense R.P. and Pender M.J. Liquefaction characteristics of crushable pumice sand, Proceedings of the 18th International conference on soil mechanics and geotechnical engineering, Paris, 2013.

Orense R.P, Pender M.J and O'Sullivan A. Liquefaction Characteristics of Pumice Sands, EQC Project 10/589, 2012.

Seed, R.B., et. al. Recent Advances in Liquefaction Engineering and Seismic Site Response Evaluation, Proceedings of the Fourth International Conference on Recent Advances in Geotechnical Engineering and Soil Dynamics and Symposium in Honor of Professor W.D. Liam Finn, San Diego, California, March 26–31, 2001.

Wesley L.D., Meyer, V. and Pender, M.J., Cone penetrometer tests in pumice sand, NZ Geomechanics News, 1998.

Youd. T.L., Idriss, I.M. Liquefaction Resistance of Soils: Summary Report from the NCEER and 1998 NCEER/NSF Workshops on Evaluation of Liquefaction Resistance of Soils, 2001.

Seismic Performance of Soil-Foundation-Structure Systems – Chouw, Orense & Larkin (Eds)
© 2017 Taylor & Francis Group, London, ISBN 978-1-138-06251-1

Influence of foundation bearing pressure on liquefaction-induced settlement

G. Barrios, T. Larkin, R.P. Orense & N. Chouw
Department of Civil and Environmental Engineering, The University of Auckland, New Zealand

K. Uemura & K. Itoh
Department of Urban and Civil Engineering, Tokyo City University, Japan

N. Kikkawa
National Institute of Occupational Safety and Health, Tokyo, Japan

ABSTRACT: Current methodologies to assess liquefaction risk and estimate induced settlement are based on a free-field data. The possible influence of the structures founded on top is neglected. Additionally, earthquake loading is commonly considered as an undrained load. Therefore, settlement results from a post-shaking particle rearrangement due to pore-pressure dissipation and transfer of load from the liquid phase to the solid phase. Recent researches have shown that these assumptions can lead to a misleading settlement estimation. A partially-drained scenario of earthquake loading has been proposed as a more accurate approach. However, only a few laboratory tests have been conducted to study the accuracy of free-field approach and partially-drained nature of the earthquake load. This study reports preliminary results from five centrifuge tests conducted in the Japanese Institute of Occupational Safety and Health (JNIOSH). Configurations of stand-alone and two adjacent rigid blocks with different mass were studied. A harmonic load of 1 Hz and a maximum amplitude of 0.2 g (prototype scale) was applied. A free-field test was also conducted. The measurements showed a significant influence of the bearing pressure on the total settlement. Additionally, settlement showed two components (during and after the shaking) supporting the partially-drained nature of the earthquake load.

1 INTRODUCTION

Widespread liquefaction-induced damage had been observed during the last big earthquakes (e.g. Maule, Chile, 2010; Tohoku, Japan, 2011 and Canterbury earthquake sequence 2010–2011). This damage produced high economic and social costs mainly related to demolition, retrofitting and rebuilding of the affected structures. This motivated researchers to improve the understanding of soil liquefaction and its effects, such as induced settlement. Rollins and Seed (1990) documented one of the first in-situ observations of the influence of buildings on the development of excess pore water pressure. Differences between the excess pore-pressure adjacent to a building and in free-field (away from the structure) were observed. In the field of laboratory experimentation, most effort has been directed towards centrifuge studies. The use of centrifuge facilities allows researchers to have an improved control of variables compared with full scale of field tests. The use of geotechnical centrifuges to reproduce field observations was presented by Bertalot, et al. (2012). One of the first attempts to study the influence of structures on liquefaction-induced settlement was presented by Dashti, et al. (2010). The authors studied the mechanism that controls liquefaction-induced settlement in the presence of shallow footings. The inaccuracy of post-liquefaction reconsolidation methodologies to estimate settlement was highlighted. The partially-drained nature of liquefaction settlement was addressed by Madabhushi and Haigh (2012).

Finally, the research conducted by Hayden, et al. (2015) summarized observations from two tests of adjacent structures on liquefiable soil. Lower settlement was observed for adjacent structures compared with stand-alone cases. All those findings indicate the complex influence of structures on the behaviour of saturated loose sands.

Significant contributions in this field have been made during the last decade. However, further research is necessary to study the complex structure-foundation-soil system of multiple structures founded on liquefiable soil. The main aim of the present work was to study the influence of different configurations of adjacent footings, with different bearing pressures, on liquefiable soil. A total of five centrifuge tests were conducted. Tests considered free-field, stand–alone and adjacent footings configurations. A laminar box was used to reduce the boundary effects and simulate lateral soil deformation. A harmonic excitation was considered. Results, showed a considerable influence of the bearing pressure and adjacent structures in the observed settlement.

2 CENTRIFUGE MODELLING

All the tests were conducted at JNIOSH. A centrifugal acceleration of 50 g was applied. A blocking system (also known as a "Touch-down system") fixes the centrifuge dynamic platform into a vertical position. This system avoids any interaction between the movements of the shaking table and the location of the platform (Nagura, et al., 1994). Further information about JNIOSH centrifuge facilities can be found at Horii, et al. (2006).

2.1 Scale factors

Soil grain geometry and density were not scaled. This was based on two main concerns. Firstly, a particle scaling can lead to a change in the soil behaviour (e.g. sand representing gravel or even clay representing sand). Secondly, the foundation size was large enough to include a considerable quantity of particles on the contact area, minimizing the particle size effects. A pore fluid with a viscosity 50 times that of water was used to avoid inconsistencies between the scaling factor for consolidation time and that for dynamic time (Konkol, 2013). For a more extended discussion about this assumption, refer to Kutter (1995).

2.2 Soil and fluid properties

Toyoura sand was used for all the tests. The main properties of this sand are presented in Table 1. These parameters were obtained from previous work conducted at the same facilities by Toyosawa, et al. (2006).

2.3 Structural models

The main parameters considered in selecting the structural models were the footing dimensions and the bearing pressure. The dimensions of the footing were limited by the width and length of the laminar box (420 × 150 mm). Target bearing pressures were chosen based on a

Table 1. Toyoura sand properties.

Parameter	Value
ρ_{min}	$1.34 \cdot 10^3 \, [kg/m^3]$
ρ_{max}	$1.65 \cdot 10^3 \, [kg/m^3]$
ρ_s	$2.64 \cdot 10^3 \, [kg/m^3]$
e_{min}	0.61
e_{max}	0.98

148

bearing capacity analysis. An analytical study was conducted using the expression presented by Terzaghi (Eq. 1).

$$q_f = cN_c + \gamma' DN_q + 0.5\gamma' BN_\gamma \qquad (1)$$

A cohesionless soil and a foundation at the surface were considered. Therefore, only the last term of Equation 1 remains. The expression proposed by Loukidis and Salgado (2010) was used to evaluate N_γ (Eq. 2). This expression was derived using Toyoura sand.

$$N_\gamma = 2.82 \exp\left(\frac{3.64 D_r}{100\%}\right)\left(\frac{\gamma' B}{p_a}\right) \qquad (2)$$

A bearing capacity of 140 kPa was obtained. Therefore, a bearing pressure of 70 kPa and 115 kPa were considered appropriate for the tests.

2.4 Model preparation

On the base of the laminar box, a 30 mm (1.5 m prototype scale) gravel layer was placed. The gravel prevented sand from clogging the injection points. Gravel also helped to obtain a uniform fluid height during saturation. On top of the gravel layer, a medium-dense sand layer (Dr = 70%), 50 mm height (2.5 m prototype scale) was placed. On top of the medium-dense soil layer, a 170 mm height (8.5 m prototype scale) of loose Toyoura sand (Dr = 40%) was pluviated. Figure 1 shows a schematic of the soil layers.

After finishing pluviation, the specimen was saturated. Due to the high fluid viscosity a vacuum system was used to achieve an adequate saturation level. The saturation process was conducted in two stages. Firstly, trapped air in the silicon oil and soil specimen was extracted without allowing oil to enter the specimen. This process was performed over an entire day to achieve a minimum amount of trapped air. The oil is then permitted to enter the sand. The saturation process takes approximately a day and results in a water table close to surface. The soil parameters obtained are presented in Table 2. As a benchmark, a free-field test (FF) was conducted. The second test involved stand-alone structures, coined as soil-footing interaction (SFI) due to the distance between the footings ($F \approx 2B$). Third and fourth tests utilised adjacent structures, coined as footing-soil-footing interaction (FSFI) due to the short distance between the footings ($F \leq 1/2B$). FSFI-1 utilised different bearing pressures (i.e. 70 kPa and 115 kPa), while, FSFI-2 employed the same bearing pressure for both footings (70 kPa).

2.5 Instrumentation

Linear vertical displacement transducers (LVDTs) were placed at both edges of the footings and in the middle of the specimen. Accelerometers, pore-pressure transduces and earth pressure sensors were also used. The locations of the devices are presented in Figure 2. Squares correspond to accelerometers; circles correspond to pore-pressure transducers; pentagons to earth-pressure sensors and empty squares to linear displacement transducers.

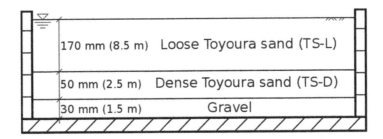

Figure 1. Soil specimen cross section.

Table 2. Soil properties for each test.

Test	Layer	Description	Height [m]	Density Dry [kg/m³]	Density Saturated [kg/m³]	e	Dr [%]
FF (1)	Top	Loose sand	8.5	$1.45 \cdot 10^3$	$1.91 \cdot 10^3$	0.830	39.5
	Middle	Dense sand	2.5	$1.54 \cdot 10^3$		0.733	65.7
	Bottom	Gravel	1.5	$2.17 \cdot 10^3$		0.175	—
FF (2)	Top	Loose sand	8.5	$1.45 \cdot 10^3$	$1.92 \cdot 10^3$	0.827	40.5
	Middle	Dense sand	2.5	$1.54 \cdot 10^3$		0.702	70.5
	Bottom	Gravel	1.5	$2.17 \cdot 10^3$		0.175	—
SFI	Top	Loose sand	8.5	$1.44 \cdot 10^3$	$1.91 \cdot 10^3$	0.831	39.3
	Middle	Dense sand	2.5	$1.55 \cdot 10^3$		0.707	72.5
	Bottom	Gravel	1.5	$2.12 \cdot 10^3$		0.205	—
FSFI-1	Top	Loose sand	8.5	$1.45 \cdot 10^3$	$1.92 \cdot 10^3$	0.827	40.4
	Middle	Dense sand	2.5	$1.53 \cdot 10^3$		0.720	69.1
	Bottom	Gravel	1.5	$2.12 \cdot 10^3$		0.205	—
FSFI-2	Top	Loose sand	8.5	$1.45 \cdot 10^3$	$1.93 \cdot 10^3$	0.827	40.4
	Middle	Dense sand	2.5	$1.55 \cdot 10^3$		0.702	73.8
	Bottom	Gravel	1.5	$2.12 \cdot 10^3$		0.205	—

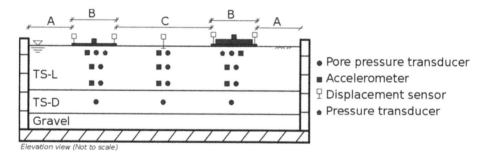

Figure 2. Location of sensors.

Table 3. Model dimensions and sensors locations.

Dim.	Description	Free-field Model (mm)	Free-field Prototype (m)	Stand-alone Model (mm)	Stand-alone Prototype (m)	Adjacent structures Model (mm)	Adjacent structures Prototype (m)
A	Distance from the edge of the foundation to the edge of the laminar box	N/A	N/A	80	4.0	125	6.25
B	Dimension of the foundation in the direction of shaking	N/A	N/A	70	3.5	70	3.5
C	Distance between footings	N/A	N/A	120	6.0	30	1.5
L	Dimension of the foundation perpendicular to the shaking	125	6.25	125	6.25	125	6.25

Table 3 summarizes the dimensions of the model and locations of sensor. Values are presented in model and prototype scales. For simplicity, the same nomenclatures (refer to Fig. 2) were used to describe all tests.

Table 4. Ground motions.

Ground motion	Frequency [Hz]		Maximum amplitude [mm]		Maximum acceleration [g]	
	Model	Prototype	Model	Prototype	Model	Prototype
1	50	1	1.0	50	10	0.2

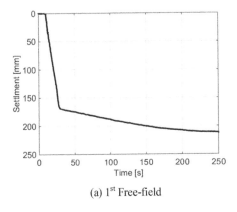

(a) 1st Free-field

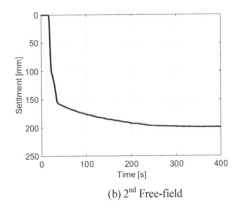

(b) 2nd Free-field

Figure 3. Total settlement recorded.

2.6 Ground motion

The ground motion applied corresponds to a harmonic signal with a frequency of 1 Hz (prototype scale). A total of 20 cycles were applied. Four cycles were used as a ramp at the beginning and the end of the motion to reach and decrease from the maximum amplitude. The other 12 cycles consisted of a constant amplitude. A maximum of 0.2 g was considered. Table 4 shows the selected properties of the ground motion.

3 FREE-FIELD (FF)

Results for the FF test are given as an average value from all the LVDTs. A second free-field test was performed to ensure the repeatability of the tests. Figure 3 shows the settlement recorded during both FF tests. The first test used a total record time of 250 s. However, by the end of the first test no stable reading was achieved from the pore pressure device. Therefore, the recording time was extended to 1000s for the following tests. Figure 4(b) shows results from the 2nd FF test. The settlement reached a stable value at approximately 400 s.

Settlement observations showed a clear co- and post-shaking settlement component. Co-shaking is defined as the settlement recorded during the shaking and post-shaking as that occurred following the shaking. Figure 4 shows the co- and post-shaking settlements in both FF tests. Very similar values of both settlement components were observed in both tests, indicating a good repeatability.

4 STAND-ALONE FOOTINGS (SFI)

Two footings with bearing pressures of 70 and 150 kPa were located a distance of about two times the footing width (see Table 3). This distance is intended to minimise the interaction between the footings. Figure 5 shows the settlement recorded at the centre of the model and beneath each foundation. A greater settlement was observed beneath both footings

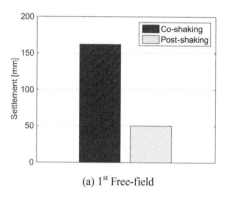
(a) 1st Free-field

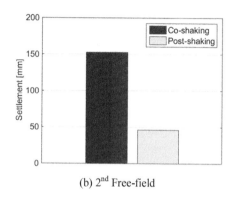
(b) 2nd Free-field

Figure 4.   Co and post-shaking settlement components.

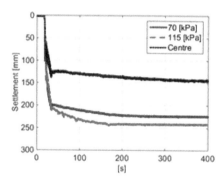

Figure 5.   Surface settlements (SFI).

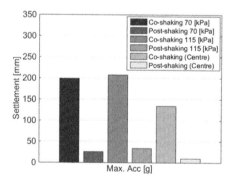

Figure 6.   Average surface settlement (SFI).

compared with the centre of the model. Also, a slightly larger settlement was recorded under the 115 kPa footing.

Figure 6 shows the co- and post-shaking settlements beneath each footing and in the centre of the model. The co- and post-shaking settlements show similar values beneath both footings. The co-shaking settlement recorded at the centre of the model was consistent with that recorded in the FF tests. This shows that the response at the centre of the model was fairly close to the FF case.

The settlement under both footings was significantly larger compared with the FF condition. However, the settlement beneath the 115 kPa footing was just slightly larger than the one for the 70 kPa footing. Therefore, the relation between the bearing pressure and

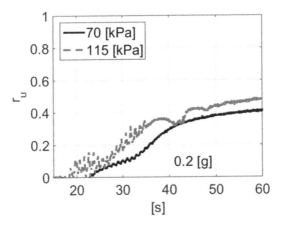

Figure 7. Pore-pressure increment under the footings (SFI).

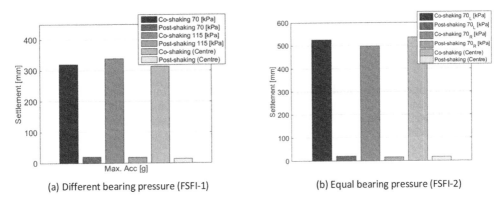

(a) Different bearing pressure (FSFI-1)   (b) Equal bearing pressure (FSFI-2)

Figure 8. Surface settlement FSFI.

settlement seems to be highly non-linear. The slightly larger pore-pressure ratio under the 115 kPa footing compared to the 70 kPa footing (Figure 7) could explain the similar settlement.

5 ADJACENT FOOTINGS (FSFI)

Firstly, the same masses considered in the SFI test (i.e. 70 and 115 kPa) were located closely adjacent to maximise the interaction. Subsequently, two masses with a bearing pressure of 70 kPa were employed utilising the same closely adjacent configuration. Figure 8 shows both components of settlement recorded in the two FSFI tests. In both tests, a similar co-shaking settlement was observed under both footings and at the centre of the model. However, a counterintuitive larger settlement was observed when the total bearing pressure (sum of both footings) was lower, i.e. 70 kPa in both footings (FSFI-2). These results can be evidence of a maximum settlement for an intermediate bearing pressure. The dilatancy tendency of soils at high confining pressures and the reduction in the pore-pressure ratio under the footing can be some of the reasons for this particular behaviour. Similar observations were previously reported by Bertalot, et al. (2013). Regarding the post-shaking settlement, beneath the footings and at the centre of the model showed similar results in both tests, however, all post-shaking settlements were lower for the equal bearing pressure test (FSFI-2).

6 CONCLUSIONS

Results from five tests conducted at JINIOSH centrifuge are presented. A 50 g centrifugal acceleration was applied for all tests. A ramped harmonic load was used. Three types of test were performed: 1) free field; 2) simulated SFI utilising independent footings (large separation distance) and 3) simulated FSFI utilising closely adjacent footings. The conclusions regarding the settlement recorded from each test are summarized below.

A clear co- and post-shaking settlement components were observed in the free-field tests. These observations reaffirm the hypothesis of a partially-drained sand behaviour under earthquake load. The repeatability of the test was verified by conducting a second FF test. The small differences found between the tests may be due to the slightly different densities achieved.

SFI was studied utilising two footings with the inter-footing distance sufficient to minimize the interaction. The recorded settlement beneath both footings was considerably larger than that recorded in the FF tests. A slightly larger settlement was observed for the footing with the larger bearing pressure (115 kPa). Therefore, the relation between the bearing pressure and settlement seems to be highly non-linear.

A clear "cluster effect" was observed in the FSFI test, i.e. the settlement values were clearly different from those recorded in the SFI test. Additionally, the highest settlement was recorded for the lowest total bearing pressure (sum of both footings, i.e. 140 kPa).

ACKNOWLEDGEMENTS

The authors also wish to thank the NZ Ministry of Business, Innovation and Employment for the financial support through the Natural Hazards Research Platform under the Award UoA 3708936. The author also wish to acknowledge the financial support of the Chilean government scholarship program "Becas Chile" for supporting his studies at the University of Auckland.

REFERENCES

Bertalot, D. y otros, 2012. *Use of centrifuge modelling to improve lessons learned from earthquake case histories.* Delft, The Netherlands, Delft University of Technology and Deltares.

Bertalot, D., Brennan, A.J. & Villalobos, F.A., 2013. Influence of bearing pressure on liquefaction-induced settlement of shallow foundations. *Geotechnique,* 63(5), pp. 391–399.

Dashti, S. y otros, 2010. Mechanisms of seismically induced settlement of buildings with shallow foundations on liquefiable soil. *Journal of geotechnical and geoenvironmental engineering,* 136(1), pp. 151–164.

Hayden, C.P. y otros, 2015. Centrifuge Tests of Adjacent Mat-Supported Buildings Affected by Liquefaction. *Journal of Geotechnical and Geoenvironmental Engineering,* 141(3), p. 04014118.

Horii, N., Itoh, K., Toyosawa, Y. & Tamate, S., 2006. *Development of the NIIS Mark-II geotechnical centrifuge.* s.l., s.n., pp. 141–146.

Konkol, J., 2013. Derivation of the scaling laws used in geotechnical centrifuge modelling application of dimensional analysis and Buckingham II theorem.

Kutter, B.L., 1995. Recent advances in centrifuge modeling of seismic shaking. *In Proc. Int. Conf. on Recent Advances in Geotechnical Earthquake Engineering and Soil Dynamics,* pp. 927–941.

Loukidis, D. & Salgado, R., 2010. Effect of relative density and stress level on the bearing capacity of footings on sands. *Purdue e-Pubs,* pp. 107–119.

Madabhushi, G.S.P. & Haigh, S.K., 2012. How well do we understand earthquake induced liquefaction? *Indian Geotechnical Journal,* 42(3), pp. 150–160.

Nagura, K., Tanaka, M., Kawasaki, K. & Higuchi, Y., 1994. *Development of an earthquake simulator for the Taisei centrifuge.* s.l., s.n., pp. 151–156.

Rolins, K.M. & Seed, H.B., 1990. Influence of builings on potential liquefaction damage. *Journal of geotechnical engineering,* 116(2), pp. 165–185.

Toyosawa, Y., Itoh, K., Surendra, T.B. & Suemasa, N., 2006. *Redistribution of active earth pressures using movable earth support apparatus in centrifuge.* London, Taylor & Francis Group, pp. 1113–1118.

Seismic Performance of Soil-Foundation-Structure Systems – Chouw, Orense & Larkin (Eds)
© 2017 Taylor & Francis Group, London, ISBN 978-1-138-06251-1

A large scale shake table test on the seismic response of a structure with SFSI and uplift

X. Qin
The University of Auckland, Auckland, New Zealand

L. Jiang
Central South University, Changsha, China

N. Chouw
The University of Auckland, Auckland, New Zealand

ABSTRACT: During an earthquake, the soil beneath the footing of a structure can deform plastically. With this nonlinear Structure-Foundation-Soil Interaction (SFSI), the response of the structure will be different from that with an assumed fixed base. Previous study of structural response with SFSI considered only the horizontal earthquake excitations. The effect of vertical component of the excitation has not been well understood. In this study a shake table test was conducted on a 1:4 scale structure model to reveal the structural response with SFSI when subjected simultaneously to horizontal and vertical excitations. To simulate the effect of SFSI the structure was placed on sand in a rigid box. The excitation used was ground acceleration recorded during the 1995 Kobe earthquake. The result showed that when only horizontal excitation was considered, the acceleration of the structure with SFSI was smaller than that with a fixed base. However, when vertical excitation was considered, the vertical acceleration of the structure with SFSI was larger than that of the structure with a fixed base.

1 INTRODUCTION

During a strong earthquake the overturning moment of a structure can exceed the available overturning resistance. A portion of the footing may intermittently separate from the supporting soil. In the event of this footing uplift, plastic deformation of the supporting soil may occur due to the concentration of stress as a result of the reduced contact interface between footing and the supporting ground. Several examples of towers and oil tanks uplifting from the underlying soil were reported after the 1952 Arvin Tehachapi, 1964 Alaska, and 1979 Imperial Valley earthquakes (Psycharis 1983).

Research in the past had shown that footing uplift can lead, in general, to a favorable reduction in maximum transverse deformation of a structure (Meek 1975). Meek concluded that this phenomenon should be considered in the design of a structure, in order to make full use of the beneficial effect of uplifts. Yim and Chopra (1985) conducted one of the early study to understand the response of flexible elastic structure with the effect of transient footing uplift. They developed a numerical model to determine the maximum acceleration of structures with SFSI during earthquake. Paolucci et al. (2008) carried out a series of shake table experiments to investigate the contact force of a rigid structure with shallow footings on sand during an earthquake. Load cells were placed at the soil-footing interface to measure the contact force. It was shown that the soil plastic deformation can reduce the contact area between soil and footing. This in turn results in a degradation of the rotation stiffness of the footing and a lengthening of the system response period.

Recently, the response of different type of structures with the effect of SFSI were studied. In the study performed by Deng et al. (2008) a series of centrifuge tests on a bridge system with plastic hinges was conducted to investigate the development of plastic hinge of a structure with SFSI. Saxena et al. (2011) numerically studied the response of nuclear reactor buildings with soil nonlinearity. Ormeño et al. (2012, 2015) investigated the effect of uplift including fluid-structure interaction. The stress development at a liquid storage tank with SFSI was discussed. Loo et al., (2012) proposed an approach to control seismic response of timber structures, by allowing the structure to rock at the same time controlling the uplift movement by using slip friction devices at its base. Patil et al. (2016) investigated the seismic response of wind turbine tower with SFSI. Qin et al. (2013) studied the effect of soil plastic deformation on the induced vibration to secondary structures.

Study of structural response with SFSI so far mainly focused on the horizontal response of structures under unidirectional horizontal excitation. Not much attention has been paid to the response of the vertical response of structure with SFSI. The effect of vertical ground motion on the structural response with SFSI has also been neglected. In this work a shake table was used to study the vertical acceleration of a large-scale structure under simultaneous horizontal and vertical excitation with SFSI.

2 SHAKE TABLE EXPERIMENT

2.1 Structure model

The experiment was conducted at the Central South University, in Changsha, China. A large-scale 6 degree-of-freedom shake table was used. The structure model was constructed based on a four-storey prototype, scaled by 4 in length. The scale factor for mass was 120. The structure model consisted of steel beams and columns with I sections. For the beam of the structure, the width of the flange was 22 mm and the depth of the web was 44 mm. For the column, the width of the flange and the depth of the web were both 35 mm. Table 1 shows the properties of the prototype and the structure.

The structure was made out of steel with a Young's modulus of 200 GPa. The total height was 3150 mm with an inter-storey height of 790 mm. The width of the footing was 1750 mm. The floor mass was 320 kg for each of the first three floors and 200 kg for the roof floor. The fundamental frequency of the structure with an assumed fixed base was 1.63 Hz.

2.2 Experimental setup

Two based conditions were considered: fixed base and sand support. When considering fixed base, the structure was fixed to the shake table. To obtain the response of the structure with the effect of SFSI, the structure was placed on sand in a box. The sand box was assumed to be rigid. Wet sand was used to fill the rigid box. The depth of the sand filled was 900 mm. The surface area of the sand was 3 m × 3 m. The setup was placed on a six degree-of-freedom shake table. Figure 1 shows the setup of the structure on sand. The shake table was capable to produce 1.8 g acceleration in the horizontal directions and 1.4 g in the vertical direction. The maximum payload of the shake table was 40000 kg.

Table 1. Properties of the prototype and structure model.

Property	Prototype	Model
Roof mass (kg)	24000	200
Floor mass (kg)	38400	320
Young's modulus (GPa)	200	200
Fundamental frequency (Hz)	1.63	1.63
Inter-storey height (m)	3.15	0.79
Base width (m)	7	1.75

Figure 1. Setup of the structure on sand.

Three accelerometers were attached at the top of one structural column to measure the acceleration of the structure in all directions (Figure 1). Laser transducer was used to measure the horizontal displacement at the top of the structure. Four laser transducers were also attached at each corner of the footing to measure the footing vertical displacement.

2.3   Ground motion

The excitation used was the ground acceleration recorded during the 1995 Kobe earthquake, at the Japan Meteorological Agency (JMA) station. The excitation was recorded in the north-south (NS), the east-west (EW), and the vertical directions. In this study, the EW component and the vertical component of the record were considered. Figure 1 shows the directions of the excitation applied to the structure. The EW component was applied to the minor axis the structure.

The ground motion was scaled by 4 to match the scale of the structure. Figure 2 shows the time histories of the excitation. The peak ground acceleration (*PGA*) of the scaled excitation in horizontal and vertical direction was 0.085 g and 0.158 g, respectively.

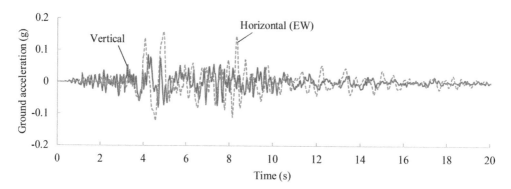

Figure 2. Horizontal and vertical components of the ground acceleration.

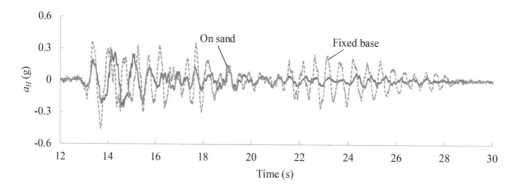

Figure 3. Horizontal acceleration at the top of the structure due to horizontal excitation.

3 EXPERIMENTAL RESULTS

3.1 *Response of the structure with SFSI due to horizontal excitation*

Figure 3 shows the horizontal acceleration (a_H) at the top of the structure in the direction of the excitation. The location of the measurement is shown in Figure 1. The result was obtained considering only the horizontal component of the excitation. The dashed and solid lines represent the structure with fixed base and on sand, respectively. As shown, the horizontal acceleration at the top of the structure on sand was smaller than that of the structure with an assumed fixed base. In the case of the structure on sand, the maximum horizontal acceleration was 0.29 g. When fixed base was considered, the maximum horizontal acceleration was 0.46 g. With the effect of SFSI, the maximum horizontal acceleration of the structure was 37% smaller.

3.2 *Response of the structure with SFSI due to vertical excitation*

Figure 4 shows the vertical acceleration (a_V) at the top of the structure due to vertical excitation. The dotted line represents the structure with an assumed fixed base and the solid line illustrates the response of the structure on sand. The frequency of the structural acceleration in the vertical direction was much higher than that of the structural acceleration in the horizontal direction. It was found that the vertical acceleration of the structure on sand was larger than that of the structure with a fixed base. The maximum vertical acceleration at the top of the structure with fixed base was 0.079 g. On the other hand, the maximum vertical acceleration of the structure on sand was 0.085 g. The vertical response of the structure considering SFSI was 8% greater than that of the structure with a fixed base.

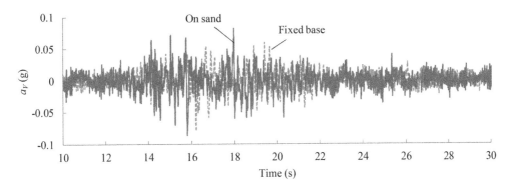

Figure 4. Vertical acceleration at the top of the structure due to vertical excitation.

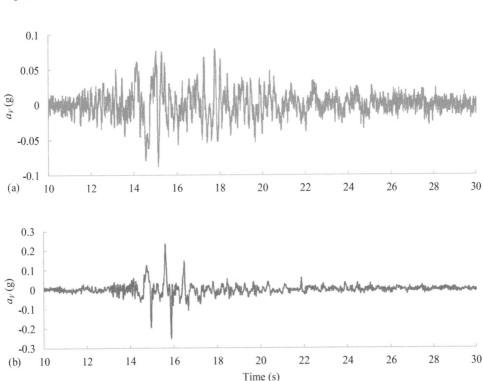

Figure 5. Vertical acceleration at the top of the structure (a) with fixed base and (b) on sand.

3.3 *Response of the structure with SFSI due to horizontal and vertical excitation*

Figure 5 shows the vertical acceleration (a_v) at the top of the structure when subjected to horizontal and vertical excitation simultaneously. While Figure 5(a) shows the case of the structure with an assumed fixed base, Figure 5(b) represents the case of the structure on sand. In the case of fixed base, the vertical acceleration of the structure due to vertical and horizontal excitations was similar to that due to vertical excitation alone. The maximum vertical acceleration of the structure was 0.086 g. In contrast, considering soil support the vertical acceleration of the structure was much larger than that due to the vertical excitation alone. Considering the simultaneous excitations, the maximum vertical acceleration at the top of the structure on sand was 0.242 g. This is because soil support causes a smaller fundamental period of the soil-foundation-structure system. The reduction of the structural period in

159

vertical direction was different from that in the horizontal direction. Thus, the effect of soil in horizontal and vertical direction were different.

4 CONCLUSIONS

This paper presents the results of a shake table experiment on a large-scale four-storey structure model with two different supports: fixed base support and sand support. The effect of SFSI on the horizontal and vertical response of the structure was investigated. The effect of soil deformation was simulated using a rigid box filled with sand. The surface area of the sand in the box was 3 m × 3 m. The structure was subjected to horizontal and vertical Kobe earthquake recorded at JMA station.

The results shown that:

- The horizontal acceleration at the top of the structure with SFSI was smaller than that of the structure with an assumed fixed base.
- For the structure on sand, the vertical acceleration was larger compared to the structure with an assumed fixed base. The finding was consistent in both cases of only the vertical excitation and simultaneous horizontal and vertical excitations.

ACKNOWLEDGEMENTS

The authors would like to acknowledge the financial support from China Railway Corporation, project number 2013G002-A-1. The authors would also like to thank the Ministry of Business, Innovation and Employment for the support through Natural Hazards Research Platform under the Award 3708936.

REFERENCES

Deng, L., Kutter, B.L., and Kunnath, S.K. 2011. Centrifuge modeling of bridge systems designed for rocking foundations. *Journal of geotechnical and geoenvironmental engineering*, 138(3): 335–344.

Loo, W.Y., Quenneville, P. and Chouw, N. 2012. A numerical study of the seismic behavior of timber shear walls with slip friction connectors *Engineering Structures* 34: 233–243.

Meek, J.W., 1975. Effects of foundation tipping on dynamic response. *Journal of the Structural Division ASCE* 101.

Ormeno, M., Larkin, T. and Chouw, N. 2012. Comparison between standards for seismic design of liquid storage tanks with respect to soil-foundation-structure interaction and uplift. *Bulletin of the New Zealand Society for Earthquake Engineering* 45(1): 40–46.

Ormeño, M., Larkin, T., and Chouw, N. 2015. Evaluation of seismic ground motion scaling procedures for linear time-history analysis of liquid storage tanks. *Engineering Structures* 102: 266–277.

Paolucci, R., Shirato, M., and Yilmaz, M.T. 2008. Seismic behaviour of shallow foundations: Shaking table experiments vs numerical modelling. *Earthquake Engineering and Structural Dynamics* 37(4): 577–595.

Patil, A., Jung, S., and Kwon, O.S. 2016. Structural performance of a parked wind turbine tower subjected to strong ground motions. *Engineering Structures*, 120, 92–102.

Psycharis, I.N. 1983. Dynamic behavior of rocking structures allowed to uplift. *Earthquake Engineering and Structural Dynamics* 11(1): 57–76.

Qin, X., Chen, Y., and Chouw, N. 2013. Effect of uplift and soil nonlinearity on plastic hinge development and induced vibrations in structures. *Advances in Structural Engineering* 16(1): 135–147.

Saxena, N., Paul, D.K., and Kumar, R. 2011. Effects of slip and separation on seismic SSI response of nuclear reactor building. *Nuclear Engineering and Design* 241(1): 12–17.

Yim, C.S. and Chopra, A.K. 1985. Simplified earthquake analysis of multistory structures with foundation uplift. *Journal of Structural Engineering ASCE*. 111(12): 2078–2731.

Seismic Performance of Soil-Foundation-Structure Systems – Chouw, Orense & Larkin (Eds)
© 2017 Taylor & Francis Group, London, ISBN 978-1-138-06251-1

Effect of ground motion characteristics on seismic response of pile foundations in liquefying soil

Andrew H.C. Chan
School of Engineering and ICT, University of Tasmania, Hobart, Australia

Xiao Y. Zhang, Liang Tang & Xian Z. Ling
School of Civil Engineering, Harbin Institute of Technology, Harbin, China

ABSTRACT: A three-dimensional Finite Element (FE) analysis is carried out to investigate the dynamic behaviour of pile foundations in liquefied soil using a fully coupled (*u-p*) formulation employed to analyze soil displacements and pore water pressures. The success of the *u-p* formulation is briefly discussed first, together with its application to various geotechnical problems. For the numerical analysis, a multi-yield-surface plasticity model is used to model the soil skeleton, and results obtained from the FE method are compared with centrifuge test data which shows excellent agreement with the observed pile and soil behaviour. The effect of the nature of the earthquake on pile performance has been studied using ten earthquake records scaled to different acceleration levels. It is found that the acceleration amplitude of the earthquake ground motion has a significant influence on the pile performance in liquefying soil. Moreover, it can be shown that the ratio of Peak Ground Velocity to Peak Ground Acceleration (PGV/PGA) of the earthquake ground motion is a better evaluation index to estimate bending moment and displacement of the pile than predominant frequency.

1 INTRODUCTION

Since its introduction by Zienkiewicz and Shiomi (1984), the so-called *u-p* formulation which utilized the displacement of the solid skeleton (*u*) and pore water pressure (*p*) as primary variables has met with great success in the numerical modelling of seismic performance of saturated soil foundation-structure systems. Due to space limitation, the formulation will not be repeated here and it can be found at Chan (1988) and Zienkiewicz et al. (1990 and 1999). The success of the procedure can be explained via Figure 1 taken from Zienkiewicz et al (1980) where π_1 and π_2 are the dimensionless permeability and excitation frequency.

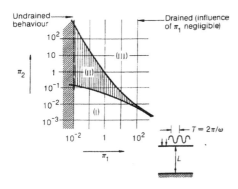

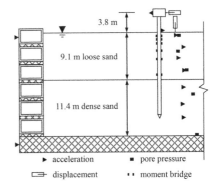

Figure 1. Validity range of various formulations (Zienkiewicz et al., 1980).

Figure 2. Centrifuge test setup, prototype scale (after Wilson, 1998).

In zone I, where the excitation frequency is low, all inertia effects can be ignored and the conventional consolidation formulation would suffice. In zone II, the relative fluid acceleration can be neglected and u-p formulation would be required while in zone III, the relative fluid acceleration is not negligible, the full u-w formulation would have to be used. For most seismic problems that would require a coupled approach, the permeability is relatively low (like, 10^{-6} m/sec) so that excess pore water pressure, once generated, will remain during the earthquake excitation period. Furthermore, unlike wave induced liquefaction problems where the consolidation formulation is sufficient, the excitation frequency is relatively high so that the inertial effect of the mixture cannot be neglected. Lastly the popularity of the u-p formulation also lies with the fact that the variables used are in common with the consolidation formulation and fluid boundary condition can be applied readily. Many successful applications of the u-p formulation has been reported in the literature. Due to space limitation, only a small selection is listed in Table 1.

Despite all these successful examples, the requirement of detailed material information has limited the applicability of the procedure to more common applications. Nevertheless it provides a useful tool, after validation, to perform detailed investigation and parametric studies so that key trends and insights can be identified. This would be otherwise very expensive and time consuming if only physical tests are to be used. In this paper, we use the u-p procedure to investigate the effect of ground motion characteristics on seismic response of pile foundations in liquefying soil and it is a good illustration of how the numerical procedure can be employed to gain a deeper understanding of the key parameters involved.

2 BACKGROUND TO THE NUMERICAL STUDY

Collapse of piled foundations in liquefiable soils have been observed in numerous strong earthquakes (Finn and Fujita, 2002). The decrease of soil strength and stiffness caused by soil liquefaction may develop large bending moments and shear forces in piles founded in liquefying soil, resulting in pile damage (Liyanapathirana and Poulos, 2005). It is well recognized that the analysis of the seismic response of piles in liquefiable ground is an extremely important and complex problem.

Ground motion characteristics are known to strongly impact the extent and nature of the interaction between soil and structure (Ghayoomi and Dashti, 2015). There remain lots of uncertainties in the seismic response of pile in liquefying ground, although a wide range of centrifuge tests (Scott J Brandenberg et al., 2005; Wilson, 1998), shaking table tests (Chang and Hutchinson, 2013; He et al., 2009), full-scale field tests (Ashford et al., 2006) and also various numerical methods (Liyanapathirana and Poulos, 2005; Rahmani and Pak, 2012) have

Table 1. Applications of the u-p formulation.

Cases	References
Dams	Elia et al 2011a and 2011b, Wu et al 2009
Embankments	Aydingun and Adalier 2003, Elia and Rouainia 2013, Oka et al 2012
Pipeline	Dunn et al 2006
Quay walls	Madabhushi et al 2008, Di and Sato 2004, Sugano et al 2000
Retaining walls	Dewoolkar et al 1999, Madabhushi and Zeng 2007, Liu et al 2014
Shallow Foundation	Coelho et al 2007, Elia and Rouainia 2014, Ghosh and Madabhushi 2004
Sheet pile walls	Viswanadham et al 2009
Tunnels	Elia et al 2013, Liu and Song 2005
Breakwaters	Ye et al 2012 and 2013
Liquefaction mitigation measures	Elgamal et al 2009
Piles	Rahman and Pak 2012, Cubrinovski et al 2008, Lu et al 2008

been employed in order to provide better insights into the seismic response of pile in liquefying soil. However, the influence of ground motion parameters [e.g., the peak ground acceleration (PGA), Arias intensity, etc] on soil-pile systems has not been investigated adequately.

In this paper, the response of a single end-bearing pile in liquefiable soil is considered by using three-dimensional FE model. Results obtained from the FE method are validated by centrifuge test data. Then, the response of pile and saturated loose sand subjected to ten ground motions are investigated. Subsequently, the effect of earthquake frequency content and PGA on pile and saturated loose sand is also examined. Finally, insights are given and conclusions drawn.

3 NUMERICAL MODELING FOR THE CENTRIFUGE EXPERIMENT

3.1 *Description of the centrifuge experiment*

As shown in Figure 2, a centrifuge test (event F in test Csp2) was performed by Wilson (1998). The soil profile consisted of two horizontal layers of saturated, fine, and uniformly graded Nevada sand in the test. The upper loose sand layer ($D_r = 35\%$) is 9.1 m thick, and the underlying dense layer ($D_r = 80\%$) is 11.4 m thick, at the prototype scale, as is the data that follows in this paragraph. Moreover, a pipe pile with a diameter of 0.67 m and wall thickness of 0.019 m is used to simulate the structural model. The embedded length of pile is about 16.8 m, and it is extended up to 3.8 m above the ground surface with a superstructure load of 500 kN.

3.2 *Finite element modelling*

All FE simulations are performed using the Open System for Earthquake Engineering Simulation, OpenSees (http://opensees.berkeley.edu, Mazzoni et al., 2009). In this research, the two-phase material response of saturated sand is based on a *u-p* formulation (Chan, 1988; Zienkiewicz et al., 1990). The FE method is selected for the spatial discretization of the equilibrium equation and Newmark implicit method is used for time integration (Katona, 1985; Katona and Zienkiewicz, 1985). The sand is modeled by a multi-yield-surface plasticity constitutive model (Elgamal et al., 2003; Parra, 1996; Yang, 2000). The pile is modeled by a bilinear material "Steel01" in OpenSees with kinematic hardening (Mazzoni et al., 2009).

The FE modeling of pile is conducted using nonlinear beam-column elements which have six degrees of freedom (DOF) at each node: three for displacements and three for rotations. Soil is modeled by an 8-node, effective-stress solid-fluid fully coupled cubic element (Biot, 1955; Lu et al., 2011). This element is based on the solid-fluid formulation for saturated soil (Biot, 1955; Chan, 1988). Each node of this element has four degrees of freedom: three for soil skeleton displacements and one for pore water pressure. The superstructure is represented by lumped mass on the pile head.

One of the significant and difficult parts of the FE simulation of pile foundations in the soil media is the connection of pile elements to the surrounding soil elements. To represent the geometric space occupied by the pile, as shown in Figure 3a, the rigid beam-column links (EI = 10,000 times the EI of the pile) are used in the direction normal to the pile vertical axis (Elgamal et al., 2008). In the soil domain 3D brick elements are connected to the pile geometric configuration at the outer nodes of these rigid links using Zerolength elements and the OpenSees EqualDOF translation constraint (Elgamal et al., 2008; Yan, 2006). Specifically, the soil node is slaved to the connection node for three translational DOFs. The connection node, at the same location as the soil node, is used to resolve the convergence problem caused by the pore water pressure DOF (Figure 3a). Moreover, the zero-length interface element (Zerolength element) is defined by two nodes (a Rigid link node and a Connection node) at the same location to allow for the slippage between pile and soil (Figure 3a). Based on Rahmani and Pak's (2012) study, an elastic-perfectly plastic material model is utilized for the interface elements. Young's modulus and yielding strain of this material are selected to be 2 MPa and 0.04, respectively.

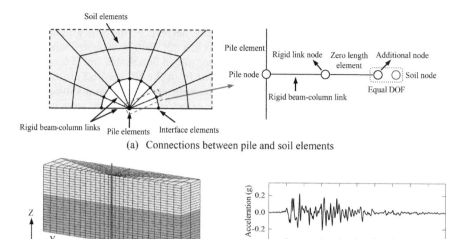

(a) Connections between pile and soil elements

(b) Finite element mesh

(c) Base input motion (Wilson, 1998)

Figure 3.   Finite element modelling.

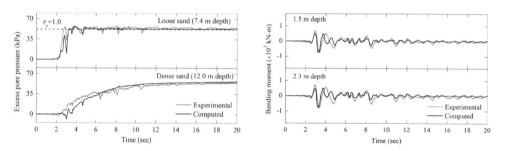

Figure 4.   Computed and experimental free-field excess pore pressure time histories.

Figure 5.   Computed and experimental bending moment time histories.

The FE mesh depicted in Figure 3b was used to model the experiment. The model is halved at the line of symmetry along the center-line of the pile (parallel with the shaking direction), and all relevant applied gravity load on the pile has been halved. The boundary conditions imposed on the FE model are: (1) at the side planes, perpendicular to the excitation direction, the nodes at any depth move together (horizontal and vertical directions); (2) the base and lateral boundaries were modeled to be impervious (due to symmetry); (3) the soil surface was stress free, with zero prescribed pore pressure; (4) the recorded base acceleration from the test (see Figure 3c) is applied to the base of the FE model as a uniform acceleration boundary condition.

3.3   *Validation of the numerical model*

The FE model is validated using the measured excess pore pressures of the sand stratum and the pile bending moments in the centrifuge test.

Prototype parameters for the soil model in the FE model are as reported in Lu et al. (2006), and permeability values of sand are based on the measured values for Nevada sand (Popescu and Prevost, 1993). The Young's modulus of the pile is 7.0×10^4 MPa, and yield moment of the pile (M_y) is 5.3×10^3 kN-m, as described in Wilson (1998). The strain hardening ratio b is 0.1 following Bhattacharya (2003).

The computed and experimental excess pore pressure time histories in the free field are presented in Figure 4. It is noted that the onset of liquefaction is attained when the excess pore pressure ratio r_u, defined as the ratio of excess pore pressure to overburden effective stress, reaches a value close to unity (Lombardi and Bhattacharya, 2016). For the loose sand, after an initial rise, the peak is achieved and then liquefaction is maintained until the end of the shaking process, while the dense sand does not liquefy. It is concluded that there is a good agreement between the computed and measured values. Figure 5 shows the computed and experimental bending moment time histories at two different depths: 1.5 and 2.3 m. The results indicate that the numerical model predicts the measured maximum bending moment well.

4 EFFECT OF GROUND MOTION CHARACTERISTICS

A total of ten earthquake ground motions (acceleration time histories) are applied to the model described in the previous section. The list of the selected input motions and their properties are presented in Table 2. The earthquake motions were selected to cover a range of intensities, durations, and frequency content. Peak ground velocity (PGV) and PGA are the maximum absolute values of velocity and acceleration. The ground motion records are obtained from the databases of the Pacific Earthquake Engineering Research Center (http://peer.berkeley.edu/). All the earthquake records are scaled to a PGA of 0.2 g, and the influence of PGA will be discussed in Section 4.3.

4.1 Free-field soil response

Liquefaction is a phenomenon that is associated with a buildup of excess pore pressure. Free-field excess pore pressure time histories of loose sand at a depth of 6 m under ten ground motions are presented in Figure 6. It is observed that in all the cases liquefaction has occurred. Moreover, the excess pore pressure distribution during ground motions with larger PGV/PGA shows sharp pore pressure decrease which is caused by soil dilation, and it will lead to an increase in the shear modulus of the soil.

4.2 Effect of earthquake frequency content on pile response

The frequency content characteristic of the ground motion is commonly reflected in predominant period, PGV/PGA and bandwidth, etc. The predominant period is generally used to estimate the seismic response of a pile (Haldar and Babu, 2010; Liyanapathirana and Poulos, 2005). The comparison between the effect of predominant period and PGV/PGA on pile response is carried out in this section, and the values of predominant period and PGV/PGA are listed in Table 2.

Table 2. Information of earthquakes used in the study.

No.	Year	Event	Station	Predominant Period (sec)	PGV/PGA (sec)
1	1992	Landers	Barstow	0.34	0.22
2	1995	Kobe	Takatori	1.22	0.21
3	1981	Westmorland	Parachute Test Site	0.42	0.21
4	1980	Irpinia	Sturno	0.20	0.18
5	1989	Loma Prieta	Treasure Island	0.96	0.16
6	1940	Imperial Valley	El Centro	0.46	0.10
7	1994	Northridge	Castatic Old Ridge Route	0.26	0.09
8	1971	San Fernando	Pacoima Dam	0.40	0.08
9	1983	Coalinga	Coalinga	0.26	0.05
10	1987	Whittier Narrows	LA-Obregon Park	0.20	0.03

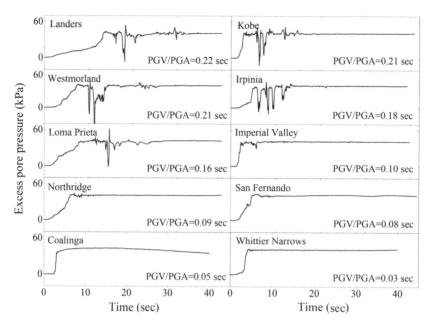

Figure 6. Free-field excess pore pressure time histories at 6 m depth under 10 earthquake records at PGA = 0.2 g (the initial effective vertical stresses for 6 m depths being 42 kPa).

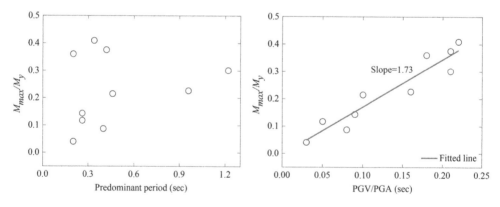

Figure 7. Variation of normalized bending moment (M_{max}/M_y) under various earthquake frequency contents.

The magnitude of maximum pile bending moment (M_{max}) and pile displacement (y_{max}) obtained from the above analysis are normalized by yield moment (M_y) and pile diameter (D), respectively. Figures 7 and 8 show the variation of normalized moment (M_{max}/M_y) and normalized displacement (y_{max}/D) under various earthquakes. It can be concluded that there is no unique relationship between the predominant period and the pile response. On the other hand, the increase in the PGV/PGA increases the normalized moment and displacement of the pile. A linear regression has been performed with the PGV/PGA values and the normalized moment and displacement (Figures 7 and 8). The slope of the fitted line of normalized moment and PGV/PGA is found to be 1.73; and 1.77 for the normalized displacement and PGV/PGA. In conclusion, the PGV/PGA is a better evaluation index to estimate bending moment and displacement of pile in liquefying ground than predominant period for the data used in this study.

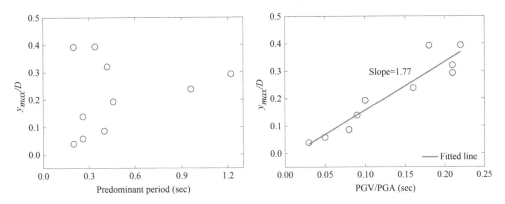

Figure 8. Variation of normalized displacement (y_{max}/D) under various earthquake frequency content.

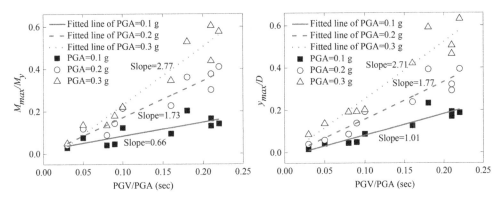

Figure 9. Variation of normalized bending moment (M_{max}/M_y) and displacement (y_{max}/D) under various PGA of earthquake ground motions.

4.3 *Effect of PGA on pile response*

All the above analyses used a PGA of 0.2 g. In this section, the significance of the PGA on a pile founded in a liquefied soil has been studied. Each earthquake has been scaled to a PGA of 0.1, 0.2, and 0.3 g peak acceleration.

The normalized moment and displacement values for three PGA values as a function of PGV/PGA are presented in Figure 9. A linear relationship between the PGV/PGA and the normalized moment and displacement of pile is also found in the case of PGA = 0.1 g and 0.3 g. The slope of the fitted line of normalized moment and PGV/PGA increases with increasing PGA, and similarly for the normalized displacement. Moreover, it is noted that the PGA has a significant effect on the dynamic response of the pile. For example, by increasing the PGA from 0.1 g to 0.3 g, the maximum bending moment and maximum lateral displacement increases about 309% and 237% respectively, when the pile is subjected to the Landers earthquake.

5 CONCLUSIONS

This paper attempts to provide better insight into the seismic performance of a pile in liquefied ground. For this, a useful 3D FE modelling method for piles in liquefiable soils using a fully coupled *u-p* formulation is employed. The success of the *u-p* formulation is briefly discussed first together with its application to various geotechnical problems. Results from a centrifuge test are simulated and the results show that the FE model has the ability to simulate pile behaviour in liquefied soil reasonably well for the case studied. Based on the FE model employed, the effects of ground motion characteristics on the dynamic behaviour of the pile foundation in liquefied ground was studied by carrying out 10 non-linear time history analyses at three intensity levels. The main conclusions derived from the present study are as follows:

1. The phenomenon of sharp excess pore pressure decreases, due to soil dilation, is seen in the earthquake ground motions with lager PGV/PGA.
2. A linear relationship between the PGV/PGA of the incident ground motion and the bending moment and displacement of the pile is found. An increase in PGV/PGA increases the bending moment and displacement of pile.
3. The maximum bending moment and lateral displacement of pile increase with the increasing PGA of the ground motion.

ACKNOWLEDGMENTS

This work was supported by the National Natural Science Foundation of China (Grant Nos. 51578195 and 51378161).

REFERENCES

Aydingun, O., Adalier, K., 2003. Numerical analysis of seismically induced liquefaction in earth embankment foundations. Part I. Benchmark model. Canadian Geotechnical Journal. 40, 753–765.

Bhattacharya, S. 2003. Pile Instability during Earthquake Liquefaction, PhD Thesis, University of Cambridge.

Biot, M.A. 1955. Theory of elasticity and consolidation for a porous anisotropic solid, Journal of Applied Physics, Vol 26, No 2, pp 182–185.

Brandenberg, S.J., Boulanger, R.W., Kutter, B.L., et al. 2005. Behaviour of pile foundations in laterally spreading ground during centrifuge tests, Journal of Geotechnical and Geoenvironmental Engineering, Vol 131, No 11, pp 1378–1391.

Chan, A.H.C. 1988. A unified finite element solution to static and dynamic geomechanics problems, PhD Thesis, University College of Swansea.

Chang, B.J., and Hutchinson, T.C. 2013. Experimental evaluation of p-y curves considering development of liquefaction, Journal of Geotechnical and Geoenvironmental Engineering, Vol 139, No 4, pp 577–586.

Coelho, F.M.G., Haigh, S.K., Madabhushi, S.P.G., 2007. Stress redistribution in liquefied ground under and around shallow foundations: experimental evidence and numerical replication. 4th International Conference on Earthquake Geotechnical Engineering.

Cubrinovski, M., Uzuoka, R., Sugita, H., Tokimatsu, K., Sato, M., Ishihara, K., Tsukamoto, Y., Kamata, T., 2008. Prediction of pile response to lateral spreading by 3-D soil-water coupled dynamic analysis: Shaking in the direction of ground flow. Soil Dynamics and Earthquake Engineering. 28, 421–435.

Dewoolkar, M.M., Pak, R.Y.S., Ko, H.Y., 1999. Centrifuge modelling of models of seismic effects on saturated earth structures. Géotechnique. 49, 247–266.

Di, Y., Sato, T., 2004. A practical numerical method for large strain liquefaction analysis of saturated soils. Soil Dynamics and Earthquake Engineering. 24, 251–260.

Dunn, S.L., Vun, P.L., Chan, A.H.C., Damgaard, J.S., 2006. Numerical modeling of wave-induced liquefaction around pipelines. Journal of Waterway, Port, Coastal, and Ocean Engineering. 132, 276–288.

Elgamal, A., Lu, J., Forcellini, D., 2009. Mitigation of liquefaction-induced lateral deformation in a sloping stratum: three-dimensional numerical simulation. Journal of Geotechnical and Geoenvironmental Engineering. 135, 1672–1682.

Elgamal, A., Yan, L., Yang, Z., et al. 2008. Three-dimensional seismic response of Humboldt Bay bridge-foundation-ground system, Journal of Structural Engineering, Vol 134, No 7, pp 1165–1176.

Elgamal, A., Yang, Z., Parra, E., et al. 2003 Modeling of cyclic mobility in saturated cohesionless soils, International Journal of Plasticity, Vol 19, No 6, pp 883–905.

Elia, G., Amorosi, A., Chan, A.H.C., Kavvadas, M.J., 2011a. Fully coupled dynamic analysis of an earth dam. Géotechnique. 61, 549–563.

Elia, G., Amorosi, A., Chan, A.H.C., Kavvadas, M.J., 2011b. Numerical Prediction of the Dynamic Behaviour of Two Earth Dams in Italy Using a Fully Coupled Nonlinear Approach. International Journal of Geomechanics. 11, 504–518.

Elia, G., Rouainia, M., 2013. Seismic performance of earth embankment using simple and advanced numerical approaches. Journal of Geotechnical and Geoenvironmental Engineering. 139, 1115–1129.

Elia, G., Rouainia, M., 2014. Performance evaluation of a shallow foundation built on structured clays under seismic loading. Bulletin of Earthquake Engineering. 12, 1537–1561.

Elia, G., Rouainia, M., Shahraki, S.N., 2013. Numerical modelling of the seismic response of tunnels observed in centrifuge experiments. 4th ECCOMAS Thematic Conference on COMPDYN, Kos Island, Greece, 12–14.

Finn, W., and Fujita, N. 2002. Piles in liquefiable soils: seismic analysis and design issues, Soil Dynamics and Earthquake Engineering, Vol 22, No 9, pp 731–742.

Ghayoomi, M., and Dashti, S. 2015. Effect of ground motion characteristics on seismic soil-foundation-structure interaction, Earthquake Spectra, Vol 31, No 3, pp 1789–1812.

Ghosh, B., Madabhushi, S.P.G., 2004. Dynamic soil structure interaction for layered and inhomogeneous ground: a comparative study. 13th World Conf on Earthquake Engineering Vancouver, B.C., Canada.

Haldar, S., and Babu, G.L.S.L.S. 2010. Failure mechanisms of pile foundations in liquefiable soil: parametric study, International Journal of Geomechanics, Vol 10, No 2, pp 74–84.

He, L., Elgamal, A., Abdoun, T., et al. 2009. Liquefaction-induced lateral load on pile in a medium Dr sand layer, Journal of Earthquake Engineering, Vol 13, No 7, pp 916–938.

Katona M.G. 1985. A general family of single-step methods for numerical time integration of structural dynamic equations, NUMETA 85, 1, 213–225.

Katona M.G. and Zienkiewicz O.C. 1985. A unified set of single step algorithms Part 3: The Beta-m method, a generalisation of the Newmark scheme, Int. J. Num. Meth. Eng., 21, 1345–1359.

Liu, H., Song, E., 2005. Seismic response of large underground structures in liquefiable soils subjected to horizontal and vertical earthquake excitations. Computers and Geotechnics. 32, 223–244.

Liu, H., Yang, G., Ling, H.I., 2014. Seismic response of multi-tiered reinforced soil retaining walls. Soil Dynamics and Earthquake Engineering. 61–62, 1–12.

Liyanapathirana, D.S., and Poulos, H.G. 2005. Seismic lateral response of piles in liquefying soil, Journal of Geotechnical and Geoenvironmental Engineering, Vol 131, No 12, pp 1466–1479.

Lombardi, D., and Bhattacharya, S. 2016. Evaluation of seismic performance of pile-supported models in liquefiable soils, Earthquake Engineering & Structural Dynamics, Vol, No, pp.

Lu, C.W., Oka, F., Zhang, F., 2008. Analysis of soil-pile-structure interaction in a two-layer ground during earthquakes considering liquefaction. International Journal for Numerical and Analytical Methods in Geomechanics. 32, 863–895.

Lu, J., Elgamal, A., Yan, L., et al. 2011. Large-scale numerical modeling in geotechnical earthquake engineering, International Journal of Geomechanics, Vol 11, No 6, pp 490–503.

Lu, J., Yang, Z., and Elgamal, A. 2006. OpenSeesPL three-dimensional lateral pile-ground interaction version 1.00 user's manual, University of California, San Diego.

Madabhushi, S.P.G., Cilingir, U., Haigh, S.K., 2008. Finite element modelling of the sismic behaviour of water front structures. The 12th International Conference of International Association for Computer Methods and Advances in Geomechanics (IACMAG), Goa, India, 1–6.

Madabhushi, S.P.G., Zeng, X., 2007. Simulating seismic response of cantilever retaining walls. Journal of Geotechnical and Geoenvironmental Engineering. 133, 539–549.

Mazzoni, S., McKenna, F., Scott, M., et al. 2009. Open system for engineering simulation user-command-language manual, version 2.0, Pacific Earthquake Engineering Research Center.

Oka, F., SongyenTsai, P., Kimoto, S., Kato, R., 2012. Numerical analysis of damage of river embankment on soft soil deposit due to earthquakes with long duration time. Proceedings of the International Symposium on Engineering Lessons Learned from the 2011 Great East Japan Earthquake, Tokyo, Japan.

Parra, E. 1996. Numerical modeling of liquefaction and lateral ground deformation including cyclic mobility and dilation response in soil systems, PhD Thesis, Rensselaer Polytechnic Institute.

Popescu, R., and Prevost, J.H. 1993. Centrifuge validation of a numerical model for dynamic soil liquefaction, Soil Dynamics and Earthquake Engineering, Vol 12, No 2, pp 73–90.

Rahmani, A., and Pak, A. 2012. Dynamic behaviour of pile foundations under cyclic loading in liquefiable soils, Computers and Geotechnics, Vol 40, No, pp 114–126.

Sugano, T., Kishitani, K., Mito, M., Nishinakagawa, K., Case, A., 2000. Shaking table tests and effective stress analyses on the dynamic behaviour of wedged caissons. 12th WCEE, Auckland, 1–8.

Viswanadham, B.V.S., Madabhushi, S.P.G., Babu, K.V., Chandrasekaran, V.S., 2009. Modelling the failure of a cantilever sheet pile wall. International Journal of Geotechnical Engineering. 3, 215–231.

Wilson, D.W. 1998. Soil-pile-superstructure interaction in liquefying sand and soft clay, PhD Thesis, University of California, Davis.

Wu, C., Ni, C., Ko, H., 2009. Seismic response of an earth dam: finite element coupling analysis and validation from centrifuge tests. Journal of Rock Mechanics and Geotechnical Engineering. 1, 56–70.

Yan, L. 2006. Sensor data analysis and information extraction for structural health monitoring, PhD Thesis, University of California, San Diego.

Yang, Z. 2000. Numerical modeling of earthquake site response including dilation and liquefaction, PhD Thesis, Columbia University.

Ye JianHong, Jeng DongSheng and Chan A.H.C., 2012 "Consolidation and dynamics of 3D unsaturated porous seabed under rigid caisson breakwater loaded by hydrostatic pressure and wave" Science China Technological Sciences, Volume 55, Issue 8, pp 2362–2376.

Ye Jianhong, Jeng Dongsheng, P.L.-F. Liu, A.H.C. Chan, Wang Rena, Zhu Changqi, 2013, Breaking wave-induced response of composite breakwater and liquefaction in seabed foundation, Coastal Engineering, 85:72–86.

Zienkiewicz, O., Chan, A., Pastor, M., et al. 1990. Static and dynamic behaviour of soils: a rational approach to quantitative solutions. I. Fully saturated problems, Proceedings of the Royal Society of London. A. Mathematical and Physical Sciences, Vol 429, No 1877, pp 285–309.

Zienkiewicz O.C. and Shiomi T. 1984. Dynamic Behaviour of saturated porous media: The generalized Biot formulation and its numerical solution, Int. J. Num. Anal. Geomech., 8, 71–96.

Zienkiewicz O.C., Chan A.H.C., Pastor M., Schrefler B.A. and Shiomi T., 1999, "Computational Geomechanics with special reference to Earthquake Engineering", John Wiley and Sons Ltd, Chichester, February.

Zienkiewicz O.C., Chang C.T. and Bettess P. 1980. Drained, undrained, consolidating and dynamic behaviour assumptions in soils, Géotechnique, 30, No. 4, 385–395.

Seismic Performance of Soil-Foundation-Structure Systems – Chouw, Orense & Larkin (Eds)
© 2017 Taylor & Francis Group, London, ISBN 978-1-138-06251-1

Seismic demand on piles in sites prone to liquefaction-induced lateral spreading

C. Barrueto
Department of Structural and Geotechnical Engineering, Pontificia Universidad Católica de Chile, Santiago, Chile

E. Sáez & C. Ledezma
Department of Structural and Geotechnical Engineering, Pontificia Universidad Católica de Chile, Santiago, Chile
Center for Integrated Natural Disaster Management CONICYT/FONDAP/15110017, Chile

ABSTRACT: Lateral spreading is one of the most important effects of liquefaction because it can cause significant ground deformation and damage to existing infrastructure. This paper studies the effects of this phenomenon in Lo Rojas, a fishing port in Coronel, southern Chile, due to the 2010 Mw 8.8 Maule earthquake using a finite element model. The mechanical characterization of the soil layers at the site was performed by laboratory tests of the materials extracted during the exploration phase, including monotonic and cyclic triaxial tests, and resonant column experiments. With the obtained laboratory curves and literature data, constitutive models for each soil layer were calibrated and used on a finite-element model in Plaxis® software. To properly reproduce the experimental behavior of the liquefiable soil layer, the UBC3D-PLM model was used. Results of the FEM model reasonably reproduce the recorded ground displacements. The seismic demand on the piles is contrasted against analytical methods, obtaining comparable results.

1 INTRODUCTION

The Mw 8.8 earthquake of 2010 affected Chilean infrastructure, causing significant destruction in the central zone of the country. Many buildings and ports near the epicenter zone were severely damaged due to ground failure and lateral spreading as described in Bray et al. (2012).

This paper studies the Lo Rojas fishering port in Coronel, Bío-Bío Region, where liquefaction-induced lateral spreading significantly damaged the existing pier. In particular, this work focuses on quantifying the seismic demand on the pier piles at Lo Rojas port. Figure 1a shows the location of the pier and the line used to measure the post-seismic residual lateral displacements, while Figure 1b displays the damaged pier due to the lateral displacement of the soil.

Several methodologies have been developed to analyze the seismic demand on piles in liquefiable soils, and to evaluate the piles response to lateral spreading. For instance, Ashford et al. (2011), provide simplified procedures to design pile foundations in soils susceptible to lateral spreading. In their work, they conclude that the dynamic response of piles with soil that has liquefied is highly nonlinear and complex. Thus the use of global nonlinear dynamic analyses can better estimate the interaction mechanisms of soil-structure interaction and how the ground's deformation pattern affects the structure performance. Finn (2005) performed an exhaustive analysis of pile response in different situations, considering the occurrence or absence of soil liquefaction, the pile head fixity condition, and the position of the liquefiable soil layer. He concluded that for piles installed in sites with shallow liquefiable soils, the bending moment is a maximum at the pile head, implying inertial interaction, and it has a large magnitude adjacent to the interface between liquefiable and non-liquefiable soil. However, despite the extensive investigations related to piles under lateral spreading,

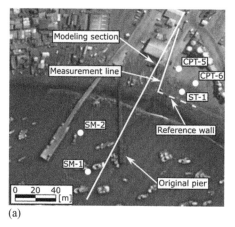 

(a)   (b)

Figure 1.   (a) Port location, lateral displacement measurement line, and modeling section line; (b) Collapsed pier due to liquefaction-induced lateral spreading.

Lombardi & Bhattacharya (2016) conclude that soil-structure interaction is still uncertain and not adequately addressed, even by modern design codes. Hence, investigations on this topic are still needed to improve the understanding of this phenomenon.

To study the lateral spreading phenomena and the seismic response of Lo Rojas pier, a FEM model was developed using the commercial software Plaxis 2D®. The liquefiable layer was modelled using two constitutive models: Hardening Soil model with Small Strain-Stiffness (HS-small), for the simulation of the static phases, and the implementation of the UBC-SAND model (Beaty & Byrne, 1998) in Plaxis® (UBC3D-PLM, Petalas & Galavi, 2013), for the simulation of the dynamic phase.

2 GEOTECHNICAL MODEL

Close to the pier site, an extensive exploration was performed to obtain soil samples and to characterize the area. The exploration included CPTu soundings, SPT boreholes, and geophysical field tests. All the collected information was used to define a geotechnical profile of the site. Details of this exploration can be found in De la Maza et al. (2017).

Figure 1a shows the section line that was modelled, while Figure 2 shows the developed geotechnical profile. As Figure 2 shows, the soil in the zone is mainly composed of four units: (H1) poorly graded sand, (H2-H3) clayey sand and high plasticity clay, (H4) low plasticity clay, and, at the left bottom part of the model, a highly cemented soil.

The information gathered from the geotechnical exploration was also used to evaluate parameters for laboratory testing. The confining pressure and the in-situ soil densities were estimated from CPTu data, while the moisture content, the specific gravity of the solids (G_s), the plasticity index (*PI*), and the grading curves of each soil were determined using soil samples from the boreholes.

3 LABORATORY TESTS

Several laboratory tests were conducted to obtain mechanical parameters for the soil layers at Lo Rojas fishermen area. All tests were performed on remolded samples with relative densities close to their estimated field condition. In the case of layers H4 and cemented soil, mechanical properties were estimated from SPT, geophysical field tests, and the literature.

H1 and H2-H3 soil layers were subjected to a set of isotropically consolidated drained (CID) and undrained (CIU) monotonic triaxial tests to obtain their friction angle (ϕ') and

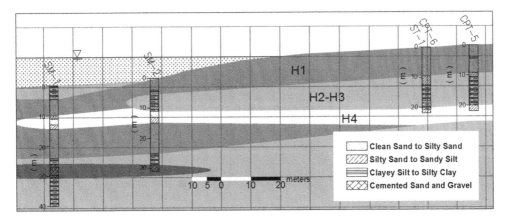

Figure 2.   Geotechnical profile at Lo Rojas, Coronel (adapted from De la Maza et al., 2017).

cohesion (c'). To characterize the cyclic behavior of these materials, i.e. to obtain the shear modulus degradation and damping curves, cyclic triaxial tests, and combined resonant column/torsional shear tests, were performed. Additionally, to define the liquefaction resistance of the liquefiable sand layer (H1), several undrained cyclic triaxial tests, with different confinement, were performed.

4   FINITE ELEMENT MODEL ON PLAXIS®

4.1   *Model parameters*

4.1.1   *Soil*
For materials H1 and H2-H3, parameters were obtained in two phases: (i) initial estimation of main parameters based on laboratory results, and (ii) triaxial tests were simulated using the Soil Test complement of Plaxis® and secondary parameters were calibrated until a satisfactory fit against laboratory strain-stress paths was obtained.

In the case of H4 soil, the mechanical parameters were obtained from correlations (Kulhawy & Mayne, 1990), shear velocity profiles calculated by De la Maza et al. (2017), and from damping and degradation curves from Vucetic & Dobry (1991). Because the cemented soil layer information is poor, it was assumed as a soft rock or gravel and it was modelled using the Mohr-Coulomb constitutive model.

For the liquefiable soil (H1), two sets of parameters were calibrated: Hardening Soil model with Small Strain-Stiffness (HS-small), and UBC3D-PLM model. This last model needs the determination of two factors related to the densification rule (fac_{hard}) and to the post-liquefaction stiffness degradation of the soil (fac_{post}). These parameters were obtained by the adjustment of liquefaction resistance curves of the model to coincide with the laboratory results. For the deepest H1 soil layer the same parameters as the shallow layer were used, but the $(N_1)_{60}$ parameter was selected according to the SPT soundings. The calibration results for layers H1 and H2-H3 can be seen, respectively, in Figures 3 and 4.

As Figure 3a shows, H1 soil behavior is well reproduced only for the contractive behavior of the material. The observed differences may be due to the inability of the constitutive model to reproduce dilatancy, which gives lower deviatoric stress and pore pressure as the strain level increases. In the case of the liquefaction resistance curve (Fig. 3b), a good fit is shown for cyclic stress ratios (CSR) less than 0.16, however, noticeable differences for higher values are observed. These differences may lead to discrepancies for the low intensity portion of the earthquake, but we believe they will not be so important for the cumulative lateral displacement induced by the motion as the simulation involves seismic records with a great number of intense loading cycles which will likely induce liquefaction. Finally, although the

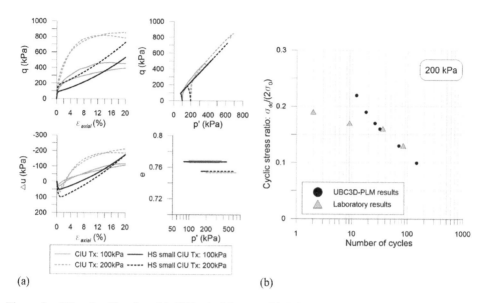

Figure 3. H1 soil calibration: (a) CIU triaxial tests with HS small model; (b) Liquefaction resistance curve of calibrated UBC3D-PLM model versus laboratory tests results at 200 kPa of effective confinement.

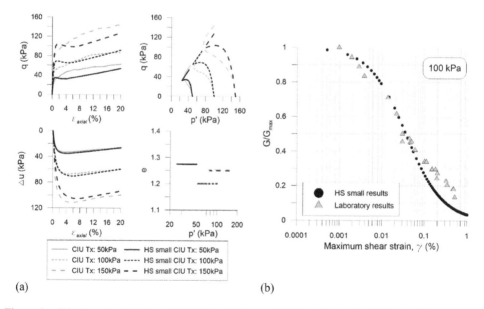

Figure 4. H2-H3 soil calibration: (a) CIU triaxial tests with HS small model; (b) Degradation curve.

static behavior is not exactly reproduced, liquefaction simulation results are comparable to laboratory data and similar deformation levels for seismic analysis of liquefiable layer are expected. Accordingly, this may lead to realistic stress levels on the pier structure during the earthquake simulation.

For the H2-H3 soil, the behavior is reasonably recreated by the HS small model, as Figures 4a and 4b show. Calibrated parameters for this material are similar to laboratory curves up to 0.06% of shear strain. Laboratory tests were conducted with confinement pressures of 50 to 150 kPa, representative of the soil pressures on the embedded piles.

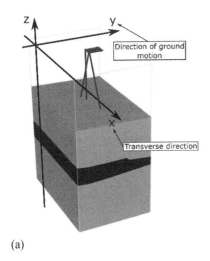 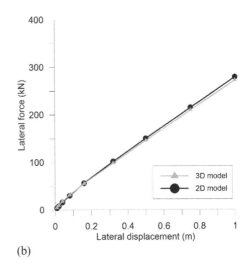

(a) (b)

Figure 5. (a) Transversal 3D model of pier for a line of piles; (b) 3D to 2D adjustment results.

4.1.2 *Pier structure*

The pier structure consisted of steel pipe piles supporting a concrete slab 200 mm thick. The port was mainly composed of two sections with different widths and transverse pile spacing. Modeling parameters were selected based on structural specifications of the original project.

Rows of embedded piles and plate elements were used to represent the longitudinal section of the port. These elements were simulated with linear-elastic models, and the interaction between the piles and the slab was considered rigid, transferring bending moments as well as shear and normal forces. As the elements representing the row of embedded piles need a maximum axial shaft resistance and a maximum base resistance for each pile, these values were calculated using average values of SPT blow counts for each soil layer and the Aoki and Velloso method (Salgado & Lee, 1999).

Equivalent 2D flexural parameters must be carefully chosen to properly represent the actual 3D behavior of the pile supported pier. A 3D model on Plaxis® was used to iteratively calibrate the diameter and thickness of the equivalent 2D embedded pile row elements to obtain a similar behavior between the 2D and the 3D force-deformation curves of each line of piles. In the 3D model the soil stratigraphy was extruded from the 2D model.

Figure 5a displays the three-dimensional FEM model used to calibrate the response of a transverse section of pier to lateral displacements, while Figure 5b shows the calibration results for one pile. In this case, to properly reproduce the 3D behavior, the diameter and thickness of the equivalent embedded beams in the 2D model had to be increased, respectively, by almost 60% and 30% with respect to the actual values, depending on the transverse section.

4.2 *Model*

To analyze the dynamic response of the Lo Rojas port to the 2010 Maule Mw 8.8 earthquake, two models were developed. First, a model without the structure was utilized to obtain a reference estimation of soil response to cyclic loading, and to verify the ability of the model to reproduce field measurements. Second, a model that has the same geotechnical characteristics of the first model but with the pier structure included. In both cases, the NS component of the ground motion recorded at the Rapel station was selected. This record was used because the distance from Lo Rojas site to the interplate fault is similar to the distance from the Rapel station to that fault plane (see De la Maza et al., 2017 for details).

4.2.1 *Model without port structure*

The model involves three major calculation phases:

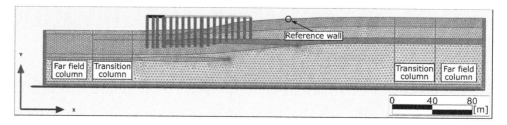

Figure 6. Finite element mesh used in Plaxis 2D® to model Lo Rojas location including the pier.

- Initial phase: Initialization of stresses. This stage was simulated with gravity loading to ensure stress equilibrium in the model.
- Second phase: This phase includes the dynamic loading, and it has the same duration as the seismic record.
- Third phase: To ensure the dissipation of the post event excess pore pressures, a consolidation calculation was simulated. This phase has a simulated duration of one day.

Boundary conditions depend on the calculation phase. For the first and third phases, boundary conditions consisted on restrained movement in the normal direction at the boundary. For the second phase (dynamic), free-field elements (at the lateral limits of the model) and a compliant base (at the bottom of the model) were used.

Free-field (non-reflecting) boundaries are applied to incorporate the propagation of waves into the far-field. This effect is incorporated by placing normal and tangential dashpots at each node of the lateral boundaries, where the parameters are selected from the soil closest to each dashpot. Compliant base boundary is designed to obtain a minimum reflection of waves at the base, and to input the ground motion.

Because the stiffness of the dynamic boundaries is related to the adjacent soil properties at the beginning of the earthquake, those borders are not strong enough to fully contain the liquefied soil layer during the seismic movement. To avoid this effect, the geotechnical profile used to create the model had to be modified at the lateral boundaries. Two soil columns were added at each side: (i) 40 m wide inelastic soil column as a transition to free field with the same properties of the original model, and (ii) 50 m wide columns composed of soil modeled with HS-small to represent the non-liquefiable far-field soil (Fig. 6).

The size of the mesh elements was selected according to Laera & Brinkgreve (2015), where it is recommended that the average size cannot be greater than one-eighth of the wavelength associated with the maximum frequency with significant energy content of the seismic signal. From the Fourier amplitude spectrum, the greatest frequency with significant energy content was around 10 Hz, and from the geophysical field tests the lowest shear wave velocity was between 120 and 130 m/s. Using this data, a maximum average size of 1.5 to 1.6 m was selected for the elements of the model. The model without the pier structure is composed of 11,633 triangular 15-node elements, and it has an average element size of 1.6 m.

4.2.2 *Model with port structure*

In this model, the boundaries and geotechnical materials are those of the previous model. To incorporate the pier, a row of embedded piles and plate elements were used. The calculation phases for this simulation are the same as the original model, but a calculation stage with plastic soil properties is added between the initial and the dynamic phases to include the initial stresses generated by the pier structure. The other phases remain identical, but in the dynamic and consolidation analysis the structure is also activated. The generated finite-element mesh is shown in Figure 6. It is composed of 11,810 elements with an average size of 1.6 m.

5 RESULTS

5.1 *FEM model displacements and structure deformation*

A post-earthquake survey (Bray et al., 2012) determined cumulative ground displacement of about 2.8 m across a 90 m line next to the pier (see Fig. 1a). Computed horizontal relative displacements across the measurement line are shown in Figure 7. It can be seen that the computed maximum horizontal displacements are similar to those measured during the post-earthquake survey. The cumulative lateral movement across the measurement line is about 2.8 m in both models. As expected, due to the pile-pinning effect, when piles are included in the model lateral displacement tend to diminish.

Simulation results show a variable lateral deformation rate (Fig. 7). The computed deformation rate in the 40 m closest to the wall face is 10 times larger than that of more distant points. In addition, close to the wall face, the measured deformation is similar to those computed by the model. The general tendency is reasonably reproduced and we believe that the model can provide reasonably realistic estimates of internal forces and displacement demand on the piles.

Due to seismic amplification, the peak accelerations increase from 0.19 g at the model base to 0.20 to 0.50 g at the model surface. Because there is no earthquake record taken close to the study site with similar geotechnical conditions, those values cannot be directly compared to the accelerations recorded at other sites. The highest PGA values occur at 30 m to 40 m from the pier, and they occur prior to liquefaction of the shallow layer.

The sensitivity of the results to the selected input motion is under study. Nevertheless, as De la Maza et al. (2017) show, the computed lateral spreading, when using the Rapel record without a pier, are approximately equal to the mean value when other available rock input motions from the Maule 2010 event are considered. We believe that a similar tendency will be found when the pier is included in the model.

Figure 8 shows the horizontal displacement contours. The accumulated horizontal displacements at the end of the earthquake are concentrated at the side of the port closer to the shore, with a maximum lateral deformation of 3.5 m. As it can be seen in the figure, and in the following one, there is a volume of soil that tends to move to the ocean and it pushes the first half of the structure with it. Post-earthquake horizontal displacements are around 1 m to 1.7 m at the ground surface in the pier area, while at the bottom part of the structure they are approximately 0.1 m to 0.5 m. Additionally, as shallow liquefied material moves more than the deep soils, significant bending moment is induced in the pile elements (Fig. 10).

As mentioned before, the pier was composed of piles and a concrete slab (divided into three sections). Horizontal displacements of the pier slabs varied between 0.35 m to 1.4 m,

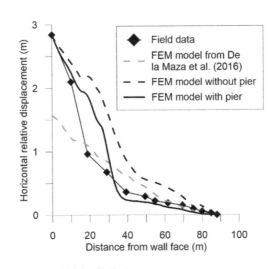

Figure 7. Model results of superficial soil displacement.

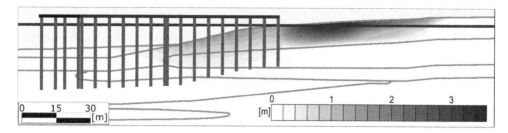

Figure 8.   Soil horizontal displacements.

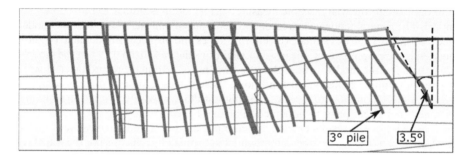

Figure 9.   Post-seismic deformation of the pier (augmented 10 times).

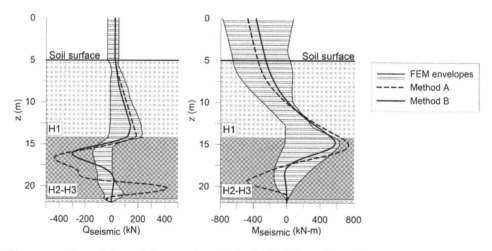

Figure 10.   Dynamic internal force envelops for the third pile from the seashore.

while vertical settlements varied between 0.01 m to 0.2 m. The maximum deformations were obtained in the part of the structure located at the steepest ground surface. At this place, the first pile of the pier has a rotational component of about 3.5° (Fig. 9).

The post-earthquake survey at Lo Rojas (Bray et al., 2012) describes the deformed structure shape as the landward part moved to the ocean compressing the pier against the seaward end. This structural response caused the seaward end to "raise" with respect to the rest of the pier. As Figure 9 shows, model results have the same qualitative deformed shape at the end of the seismic motion. However, the model is unable to capture the interaction of piles once inter-pile contact is made as the deformation increases.

5.2 *Pile stress analysis*

Figure 10 shows envelops of instantaneous seismically induced internal shear ($Q_{seismic}$) and bending moment ($M_{seismic}$) profiles of the third pile when moving from the seashore to the ocean. In the case of shear forces, the diagrams show the effect of the shallow liquefiable soil thrust pushing to left, while close to end of the pile, a reaction equilibrates this lateral force. Regarding the bending moment diagram, there are three critical sections with similar high values: (i) the slab-pile connection, (ii) below the ground level in the liquefied layer, and (iii) close to the interface between the shallow material and the non-liquefiable layer. Those zones are critical in terms of structural design, as the first one (slab-pile connection) was the location of several structural failures observed after the earthquake. The nominal yield bending moment of the piles is about 142 kN-m. This value is less than the maximum computed moment, indicating that the piles had already reached the yield in the FEM model.

Shear and bending moment diagrams obtained from the FEM model were compared against two simplified methods to calculate the lateral spreading effects on the piles (Fig. 10). The methodologies and results of these approximated models are described below.

5.2.1 *Method A: FEM model without pier and LPile® simulation*

This approximation consisted in the simulation of a single pile using the LPile® software, under an imposed displacement profile. The software solves the resulting differential equation for a beam-column element using p-y curves. The lateral displacement profile used in this analysis was extracted from the results of the FEM model without the pier structure at the location of the analyzed pile (i.e. third pile out from the seashore).

The geometry and properties of the pile in LPile® were defined by the project specifications and it was modeled as linear-elastic element. To include the inertial effect of the concrete slab, a shear force, equal to the product of the tributary mass, estimated as the superstructure mass over one pile without service loads, and the slab acceleration, was added at the top of the pile. Because there is no information about ground acceleration at the site analyzed, a value of 0.4 g was chosen to calculate the imposed shear force, using as reference the PGA of the 2010 Maule Earthquake recorded at Concepción city (http://terremotos.ing.uchile.cl/).

Soil was characterized by in-situ and laboratory information, specifically, soil type, friction angle (ϕ'), cohesion (c') and effective specific weight (γ'). This data was required by the software to estimate the p-y curves.

5.2.2 *Method B: Slide® software analysis and LPile® model*

This simplified methodology, based on Ashford et al. (2011) and MCEER/ATC-49-1 (ATC/MCEER Joint Venture 2003), was implemented using LPile® and Slide® softwares. The main steps of this method are:

- Classify and assign properties to soil using in-situ and laboratory information. For liquefiable layers, S_{ur} was assigned using Ledezma & Bray (2010).
- Perform a pseudo-static stability analysis with different horizontal acceleration values (k_h) to calculate the restitutive force that ensures a safety factor (*FS*) of 1. This supporting force must be located at ground surface of the vertical component of the center of gravity of the potentially sliding mass, and it is calculated performing a back analysis of failure surfaces until the requirement of *FS* = 1 is reached.
- Estimate the lateral displacement due to the seismic demand using the Bray & Travasarou (2007) formula for each k_h value used in the previous phase.
- Obtain the restraining forces from the structure using a simplified model of the pile-soil system. Perform a pseudo-static pushover analysis over a pile for incremental soil displacements assuming the displacement profile shown in Figure 11a.
- Obtain the curves of restitutive forces versus displacements, from steps one and two, and the restraining structure force versus displacement, from the third step.
- Obtain the intersection of the curves described in the previous step (force and displacement), i.e. ensure displacement compatibility.
- Impose the resultant displacement over the pile to obtain the internal forces acting on it.

179

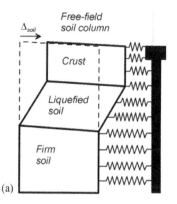 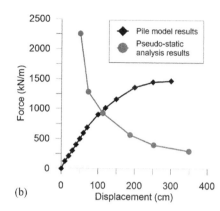

(a)   (b)

Figure 11. (a) Imposed displacement profile over pile in pushover analysis, modified from Ashford et al. (2011); (b) Force-displacement curves used in the method B.

The Figure 11b shows the curves obtained for Method B. As this is a 2D plain-strain analysis, the results need to be modified to compare them against a single pile model. To use the curves together, the results from the single pile model need to be multiplied by the number of piles contained in the expected failure surface and divided by the transversal pile spacing. This procedure introduces several assumptions as the structure may have different transverse pile spacing, piles are not likely to cross the failure surface at the same level, and/or they may have a 3D orientation. Thus results are only an approximation of the actual soil-structure interaction and it gives a first approximation of the piles performance due to lateral spreading.

5.2.3 Simplified methods results

As Figure 10 shows, the curves obtained from methods A and B are, in general, contained by the FEM model envelopes. The shapes of the curves are relatively similar and suggest agreement with the assumption of liquefied soil behavior. However, they have different maximum values and these occur at different locations. In the case of the shear force, method A predicts a maximum value of more than twice that of the FEM model, and it takes place at a different location. These values are probably produced because this strategy overestimates the deformations of the non-liquefiable soil next to the analyzed pile, as it does not account for the displacement restriction imposed by the pier.

As Figure 10 shows, bending moments obtained by the simplified methodologies are approximately a half of the FEM results close to the pile head. These dissimilarities occur because the simplified methods do not properly capture the inertial interaction between the piles and the superstructure. In the case of simplified methods, inertial effects are taken as a boundary condition with a low shear force at the pile head, which leads to low bending moments. Whereas in the FEM model, the bending moment is modelled as a fixed connection between the pier slab and the piles,

Finally, although method B gives smaller values than method A, it has the significant advantage that it does not need a sophisticated model to give results.

6 CONCLUSIONS

The main conclusions of this study are:

– UBC3D-PLM is able to represent, to a useful degree, the seismic soil response of liquefiable materials. The horizontal relative displacements predicted by the FEM model are greater than the field observations at points close to the reference wall. This could be related to the soil layer simulated using UBC3D-PLM. This layer liquefies earlier in the model than it does in the experiments, hence the post-liquefied behavior predicted by the model is less rigid than the actual behavior of soil.

- Simplified methods to assess lateral spreading effects over piles are a good way to obtain a first approximation of the structural response of piles. Although they do not reach the same maximum values of internal forces, they have similar shapes. These methods can give a first estimate of the deformation and stresses on the structure.
- Due to the 3D nature of the port site, results of a 2D model are only an approximation. A two dimensional model enforces a plane-strain condition modifying the loading transfer between the piles and the surrounding soil, which does not include three dimensional topography/bathymetry and soil variability, and it cannot simulate the complete three-component, i.e. out of plane, seismic loading. More realistic results could be, in principle, be achieved with a 3D model incorporating all the features mentioned above. Nevertheless, simulated residual 2D deformations of the pier, obtained with the benefit of post-event field measurements, are very similar to the values measured during the post-earthquake survey.

REFERENCES

Ashford, S.A., Boulanger, R.W. & Brandenberg, S.J. 2011. Recommended design practice for pile foundations in laterally spreading ground. PEER report 2011/04, Pacific Earthquake Engineering Research Center. University of California, Berkeley.

ATC/MCEER Joint Venture 2003. Recommended LRFD guidelines for the seismic design of highway bridges. Liquefaction Study Report No. MCEER/ATC-49–1 Prepared under NCHRP Project 12–49, Applied Technology Council, Multidisciplinary Center for Earthquake Engineering Research. Buffalo, N.Y.

Bardet, J.P. 1997. *Experimental soil mechanics*. Prentice Hall.

Beaty, M. & Byrne, P. 1998. An effective stress model for predicting liquefaction behavior of sand. *Geotechnical Earthquake Engineering and Soil Dynamics III ASCE Geotechnical Special Publication No 75*, **1**:766–777.

Boulanger, R.W., Chang, D., Gulerce, U., Brandenberg, S.J. & Kutter, B.L. 2006. Evaluating pile pinning effects on abutments over liquefied ground. In: Seismic Performance and Simulation of Pile Foundations in Liquefied and Laterally Spreading Ground (pp. 306–318). ASCE.

Bray, J., Rollins, K., Hutchinson, T., Verdugo, R., Ledezma, C., Mylonakis, G., Assimaki, D., Montalva, G., Arduino, P., Olson, S.M., Kayen, R., Hashash, Y. & Candia, G. 2012. Effects of ground failure on buildings, ports, and industrial facilities. *Earthquake Spectra*, **28** (S1), S97-S118.

Bray, J., Travasarou, T. 2007. Simplified procedure for estimating earthquake-induced deviatoric slope displacements. *Journal of Geotechnical and Geoenvironmental Engineering*, **133** (4), 381–392.

De la Maza, G., Williams, N., Sáez, E., Rollins, K. & Ledezma, C. 2017. Liquefaction-induced lateral spread in Lo Rojas, Coronel, Chile. Field study and numerical modeling. *Earthquake Spectra*, **33** (1), 219–240.

Finn, W. D. L. 2005. A study of piles during earthquakes: Issues of design and analysis. *Bulletin of Earthquake Engineering*, **3**(2), 141–234.

Kulhawy, F.H. & Mayne, P.W. 1990. Manual on estimating soil properties for foundation design. Electric Power Research Institute, United States.

Laera, A. & Brinkgreve, R.B.J. 2015. Site response analysis and liquefaction evaluation. Available in the Plaxis Knowledge Base website.

Ledezma, C. & Bray, J. 2010. Probabilistic performance-based procedure to evaluate pile foundations at sites with liquefaction-induced lateral displacement. *Journal of Geotechnical and Geoenvironmental Engineering*, **136**(3): 464–476.

Lombardi, D. & Bhattacharya, S. 2016. Evaluation of seismic performance of pile-supported models in liquefiable soils. *Earthquake Engineering Structural Dynamics*, **45**, 1019–1038.

Lo Presti, D.C.F., Pedroni, S., Cavallaro, A., Jamiolkowski, M. & Pallara, O. 1997. Shear modulus and damping of soils. *Géotechnique*, **47** (3), 603–617.

Petalas, A. & Galavi, V. 2013. Plaxis liquefaction model UBC3D-PLM. Available in the Plaxis Knowledge Base website.

Plaxis 2-D. 2015. Reference Manual. Available in the Plaxis Knowledge Base website.

Salgado, R. & Lee, J. 1999. Pile design based on cone penetration test results. FHWA/IN/JTRP-99/8, Purdue University, West Lafayette, IN.

Vucetic, M. & Dobry, R. 1991. Effect of soil plasticity on cyclic response. *Journal of Geotechnical Engineering*, **117** (1), 89–107.

Seismic Performance of Soil-Foundation-Structure Systems – Chouw, Orense & Larkin (Eds)
© 2017 Taylor & Francis Group, London, ISBN 978-1-138-06251-1

Summary of discussion sessions

ABSTRACT: At the end of each day of the workshop, a discussion session was conducted with the aim of eliciting important comments about issues and new trends related to the current state of understanding of the seismic performance of Soil-Foundation-Structure (SFS) systems and related topics. The first discussion session dealt with the dynamics of SFS systems while the second session focused on nonlinear SFS interaction. During the sessions, the discussions also shifted to identifying the significant issues presented by each workshop speaker. In this summary, only the general topics taken up for discussion are outlined and presented. The more detailed topic discussions are incorporated by the authors when they revised their respective papers for publication.

SPECIAL NOTE

The two discussion sessions were conducted under intense yet enjoyable exchange of opinions among the workshop participants. Because of the inherent difficulty in capturing everything that was said, thereby perhaps losing the context when transforming them into a written form, readers are cautioned that what follows represents the best efforts of the editors to capture the flow and cut-and-thrust of the discussions.

1   DISCUSSION SESSION 1

The theme for the Discussion Session 1 was on the dynamics of soil-foundation-structure (SFS) systems. With both structural and geotechnical engineers attending the workshop, it was deemed that it would be best if some terminologies were first clarified. For example, when dealing with the coupling of geotechnical and structural engineering approaches.

The first session attempted to answer the following questions: (1) What are the primary mechanisms in the dynamics of soil-foundation-structure systems? (2) Identification of mechanisms that are not fully captured by a simplified SFS approach? (3) How can the principal mechanisms of interaction be incorporated into the design procedures?

However, during the session, the discussion shifted to an examination of individual papers. The major points that came out of the discussion are summarised below.

1.1   *Interaction between structural and geotechnical engineers*

A common language between structural and geotechnical engineers should be fostered. Structural engineers should ask geotechnical engineers about e.g. the friction angle, void ratio, cohesion, shear strength, spring and dashpots. On the other hand, the geotechnical engineer should be aware of the building tolerance with regard to settlement. In summary, engineers should be aware of the relevant questions to ask their colleagues. Engineers, who will be the future practitioners, should encourage to develop a common language to enable communication between structural and geotechnical groups. In addition, focusing on either performance-based design or capacity design will help students to first understand the fundamentals and then understand the design standards. Engineers need to be educated about mechanism that follow from the behavior of soil-foundation-structure system.

Generally, when a question is raised by structural and geotechnical engineers, a presumed mechanism is behind that question. It is better to clearly define the failure mechanism under investigation rather than, for example, requesting information on the stiffness, which is obviously related to strain level. It is better to consider the whole SFS system as a holistic entity, rather than saying a structure supported by a foundation or vice-versa. Considering the system as holistic will lead to identification of the critical component and bring together the geotechnical and structural engineers.

A development of middle ground between structural and geotechnical engineers requires collaboration. Currently, geotechnical engineers consider soil only, while structural engineers design structures somewhat independent of geotechnical behaviour. In reality, the structural response and the geotechnical behavior occur simultaneously. On a daily basis, the uncertainty in geotechnical engineering is significant.

An ideal objective would be to develop design specifications which embody the necessary part of structural and geotechnical engineering concurrently. In the US, when structural and geotechnical engineers work together, liability is still a problem and they can be sued. Consequently, everybody tries to shun away from their responsibility. However, professional indemnity is important; New Zealand is a good place to start this cooperation, because it is a small community—insurance companies can be on-board and everybody can work together.

1.2 Use of free-field motion in SFSI analyses

A question was raised on the applicability of using free-field motions as input in routine analysis of the seismic response of structures. Many simple laboratory experiments have been conducted to record the ground motions at different locations within the model ground and structure. The results showed that the recorded motions in the free field and underneath the structure are different because the system with and without a structure is not the same, i.e. there is different states of confining pressure. The experiments using a small laminar box showed that the recorded response on top of the structure cannot be replicated by dynamic analysis using free-field input motion. Such difference in response may also be due to the wavelengths compared to the foundation width. The bottom line is there are many factors affecting the response of soil-foundation-structure system.

1.3 Effects of earthquake-induced landslides and multiple seismic events on SFS systems

Sometimes, a question that needs to be answered is how to consider a sequence of strong events, such as those experienced during the 2010–2011 Christchurch earthquake and the 2016 Kumamoto earthquake. Although mitigation measures are available, earthquake-induced landslides are difficult to monitor (e.g. through an alarm systems) because there is no prior knowledge as to where the locations of the critical slopes are. Identifying vulnerable slopes by investigating subsoil condition would be very costly, both in mountain areas and in urban cities. Nevertheless, in the US, aerial photos of remnants of previous earthquake-induced landslides can be used as starting point in landslide hazard assessment. A possible next step would be to monitor these slopes in advance. However, an identification of all slopes that will fail may not be important; rather, it may be better to focus on identifying slopes that are critical to infrastructure, i.e. monitor only the slopes that will have significant effect on bridges.

How do we stop a severe earthquake-induced landslide from affecting structures, like bridges in mountainous areas? While prevention is not to be possible, a system to reveal the state of the bridge after the earthquake, say using drone or any instrumentation, is possible. In order to minimise the effect of landslides on transport infrastructure, redundancy of the system will help.

Based on analyses of houses following the 2016 Kumamoto earthquake sequence, a 50% increase in strength is required to resist the two sequential events. However, when dealing with the effect of multiple events, there is a need to consider the relevant geotechnical issues, such as the accumulation of high excess pore water pressure or accumulated deformation of retaining walls; every earthquake event could push the retaining structures until they reach the collapse mechanism.

1.4 *SFS system response to multiple simultaneous hazards*

In 2016, Wellington experienced a big earthquake and heavy rainfall almost simultaneously. However, in practice, structures are not usually designed to withstand both extreme events because of the cost involved. Commonly, the cost controls and the optimisation of expenditure is a problem. In Japan, following the 2011 earthquake, they increased the design spectrum gradually because the engineers were certain that stronger shaking in the future would happen.

The involvement of the contractors at various stages of the project is worthwhile; involving them in the discussion would not always result in the cheapest solution, but ideally trusting the contractors is recommended. However, cases where the contractors were remiss were discussed; for example, crosshole testing in Christchurch between the areas where low mobility grout remediation was done showed the ground actually became worse. Although this technique was one of the most recommended, the contractor failed to make the appropriate checks. It is therefore necessary for engineers to get involved in the construction phase and monitor the implementation of design.

1.5 *Non-structural components*

Currently, it is quite difficult to convince people that secondary components may have an impact on structural response. Even in weak earthquakes, secondary components could damage the primary structure. It is imperative to inform clients about the potential of the secondary structure to cause substantial damage to the main structure.

1.6 *Comments on damping*

The use of hammer tests to identify damping was proposed. This may be possible, although there is difficulty in placing the load and in instrumenting the system. Damping is affected more by strain, rather than by stiffness. From a structural point of view, damping may be defined in relation to plastic energy dissipation and friction between members. For bridges (as well as shallow foundations), which incorporate pounding, the combined vertical and rocking motions will induce significant stresses.

2 DISCUSSION SESSION 2

The intended topics for the second discussion session was nonlinear soil-foundation-structure interaction (SFSI). The initial discussion focused on terminology, i.e. SSI (soil-structure interaction) vs. SFSI (soil-foundation-structure interaction). The stiffness of the foundation ground may affect significantly the period/frequency of the structure. Since elastic behaviour of the soil occurs under very small strain, nonlinear interaction between soil and foundation and between foundation and structure, referred to as SFSI, is very important.

There were three main questions raised for discussion: (1) how to deal with uncertainty in SSI/SFSI; (2) the calculation of the energy balance in SSI/SFSI; and (3) how to treat nonlinearity in structures and/or soil-foundation interaction (including geometric and material nonlinearities). Below is a summary of the resulting discussions in response to the above questions.

2.1 *Uncertainties in SSI/SFSI*

There are three main sources of uncertainty: structural properties, soil and foundation properties, and earthquake motions. Generally, shallow foundations can uplift, resulting in elongation of the period of the response. While rocking may be beneficial, there is a need to consider the resulting tilt; for example, different soil conditions may result in different magnitudes of residual tilt. There are uncertainties regarding soil properties, although an estimate can be obtained through site/ground investigation. However, borehole data can be misleading, and the coefficient of variation (COV) is usually taken as 0.30; such approach is incorrect. If one goes to a site and does multiple boreholes, different shear wave velocity profiles are usually obtain. In terms of site investigation, geophysical methods may provide good quality data.

The previous workshop (held in 2009) discussed how to deal with variable soil; e.g. one possible way is to ascertain the range of soil properties and carry out a sensitivity analysis, using different ground motions, to see the variability of the ground response.

Earthquake motions based on design codes have a significant effect on SSI resulting in a large variability in response. Recorded ground motions, from e.g. PEER records, have different magnitudes, focal depths, source mechanism and different geometric relationship to the site. All these factors contribute to the large uncertainly of the ground motions.

The wavelength of ground motion is important, as well as the signature of the site. To carry out time history analysis one could take appropriate earthquake motions, calculate the COV and draw the response spectra.

Some of the perceived uncertainties depend on the ground motion parameters. In some standards, in order to deal with uncertainty at least in terms of the soil, the stiffness of the soil to be used in the calculations is set as between half or double the value in the design and then a check is made to assess the sensitivity of the structural response. If a thin silt continuous soft soil layer is present, local enhancement of shear stress and displacement may induce a failure surface, that encompasses the entire foundation. Thus, the strain level is an indicator for the state of soil stress.

There is incompatibility between the desire to have an adequate designed and constructed facilities and what is achievable in reality. Often, limitation in resources, i.e. time and budget, does not permit the realization even if the knowledge required is available. While it is not yet clear how SFSI will change during ground excitation, the key issues are in the incorporation of appropriate earthquake ground motion and SFS system properties in the modeling. An attempt to compartmentalize the damage to SFS system, i.e. a tribute failure either to the geotechnical aspect or the structure aspect, is not helpful, since it does not reinforce that this SFS system is fully integrated. For example, in the case liquefied soil the structural elements can remain damage free, even though the whole structure undergoes unacceptable global displacement. This means from the structural point of view, the structure is locally intact but globally distress. The important is the structure is no longer useable, therefore the SFS system is in a state of "failure".

2.2 *Consequence of liquefied soil for surface ground motion development*

In liquefied soil, shear wave cannot be transmitted, consequently, the horizontal component of the ground motion decreases. In contrast, P-waves can be transmitted and result in an amplification of the vertical component of the ground motion. In the liquefied medium the transmitted P-waves are compression waves in a "liquid" medium.

2.3 *Simulation of nonlinearity in structure and/or soil-foundation*

In addition to geometry changes such as foundation rocking, the interface between soil and foundation can also be a source of soil nonlinearity. However, such nonlinearity can be simplified in practice. However, to model interface action is a challenge and simplifications can lead to more complications.

Rocking of foundation is only beneficial to the structural elements, but not to the overall performance of the structure and the supporting ground, if the foundation ground is poor. A small footing rotation can lead to large difference in SFS system response. However, worldwide this feature is not yet at the stage when it can be codified. A good example is the South Rangitikei Rail Bridge, whose piers have been free to rock for a number of decades. Although from a human perception, rocking or residual tilting is not desirable; an acceptable tilt of 0.1 rad can be seen clearly in many structures.

It is known that the loading path generally will affect the response of SFS systems. Nonlinearity can be approximately addressed through the equivalent linear method (i.e. strain compatibility); however, there is perhaps a need to specify the circumstances that would support its use in practice, e.g. in small earthquake events. A common approach is to use equivalent linear analysis (ELA) as starting point, but after that a more sophisticated analysis is recommended. ELA appears to be the answer for moderate motions, but there are many better tools currently available.

While in Japan a pile is not designed to behave plastically, in NZ plastic hinge is allowed on piles but only at locations which cannot be seen, i.e. limited yielding is allowed underneath the soil surface. Plastic hinging is not part of high-seismic design of structures for NZ Transport Agency, and a limit to the rotation within the plastic hinges of the piles is specified.

MAJOR CONTRIBUTORS

S. Iai & H. Hao (Session 1 Chairmen), M. Pender & I. Takewaki (Session 2 Chairmen), A. Chan, C.Y. Chin, N. Chouw, P. Clayton, I. Dimitrakopoulos, M. Larisch, T. Larkin, M. Millen, R. Orense, K. Stokoe, L. Storie, J. Toh and I. Towhata.

Seismic Performance of Soil-Foundation-Structure Systems – Chouw, Orense & Larkin (Eds)
© 2017 Taylor & Francis Group, London, ISBN 978-1-138-06251-1

Author index

Printed and bound by CPI Group (UK) Ltd, Croydon, CR0 4YY

18/10/2024

01776219-0001